Transitioning to Java

Kickstart your polyglot programming journey by getting a clear understanding of Java

Ken Fogel

‹packt›

BIRMINGHAM—MUMBAI

Transitioning to Java

Copyright © 2023 Packt Publishing

Group Product Manager: Gebin George
Publishing Product Manager: Kunal Sawant
Senior Editor: Rohit Singh
Technical Editor: Pradeep Sahu
Copy Editor: Safis Editing
Project Coordinator: Prajakta Naik
Proofreader: Safis Editing
Indexer: Manju Arasan
Production Designer: Shyam Sundar Korumilli
Developer Relations Marketing Executives: Sonia Chauhan and Rayyan Khan
Business Development Executive: Kriti Sharma

First published: May 2023

Production reference: 2280423

Published by Packt Publishing Ltd.
Livery Place
35 Livery Street
Birmingham
B3 2PB, UK.

ISBN 978-1-80461-401-3

www.packtpub.com

To my wife, Cheryl Rayson, who supported me in several careers before I discovered programming and then teaching. To my children, Jessica and Jeffrey, who put up with the long hours I spent in my office preparing for my classes and learning to be a better programmer. To my grandchildren, Hunter and Isabella, who believe I have the best toys.

– Ken Fogel

Foreword

I have known and worked with Ken Fogel for many years! While I was at Sun Microsystems in the early 2000s, working on NetBeans as a tool to connect the Java community, we quickly came across Ken. Why? Because we wanted to bring NetBeans to students via their educators, their lecturers, professors, trainers, and so on. So, the very first person who got the message and ran with it was Ken. An idea we had was that if we could get educators together, they'd be able to exchange their lesson materials. Since all Java courses at some point handle the topic of tooling, we thought it would be interesting to have an exchange service for and by educators so that the course materials that they put together for their students could be put in some kind of repository that they'd all have access to.

It was one of many interesting ideas in principle that never really took off in a big way, though that did result in a few strong connections being made that would turn out to be of great value later. Ken was one of these! From the outset, it was clear that he thought globally, with a big vision beyond the specifics of the institution that was so lucky to have him in Canada. He began by starting a NetBeans conference there, which he ran for a few years, and then expanded beyond NetBeans, and he now is the driving force behind the big international Java Champions conference that is held virtually and has buy-in across the Java Champions community. Of course, that's all in the context of Ken becoming a Java Champion in the meantime. Because that's what happens when you activate yourself, enable yourself, switch yourself on to the broader Java ecosystem, and begin contributing to it and participating in it. As you take on more responsibilities and participate in more and more activities, you inspire others, which in turn rejuvenates and inspires you on your path in ways you would never have thought possible.

Another memory I have of Ken is on a stage in the Moscone Center in San Francisco on a discussion panel with another Canadian, James Gosling. Could it be my imagination, or is Ken the James Gosling of the Java education scene, in that in as much as James has inspired and initiated the Java language and platform, so Ken has inspired and blown life into the Java education scene?

With this book, he's certainly contributed massively to that endeavor. When you're transitioning to Java, there's really nothing you need to know beyond what Ken describes so comprehensively and succinctly in this book. Starting from setting up your environment to language fundamentals, then onwards to Java syntax and design principles, and from there to documentation and testing, there's no stone that Ken leaves unturned. It is admirable how seamlessly he's weaved new language constructs such as Records and Streams into the fabric of his work while also covering frontend technologies such as JavaFX and backend services incorporating Jakarta EE.

I'm very happy and proud to have been given this opportunity to put the spotlight on its author and to share with you the enthusiasm, passion, and sheer joy of programming that has always defined Ken and is an immutable part of his persona and such a wonderful person to know and to have in one's circle.

In this book, he shares all his knowledge in a structured and measured manner that specifies him as an educator and enthusiast of Java on the world stage.

Geertjan Wielenga

Senior Director of Open Source Projects, Azul

Contributors

About the author

Ken Fogel began his career as a freelance developer, as well as teaching evening courses in computer science after working in graphics and photography. After 7 years as a developer, he was invited to join the computer science faculty at Dawson College, where he taught for 31 years, serving as department chairperson for 25 of those years. He is currently a Java Champion, an organizer of the annual JChampions Conference, and a member of the Java Community Process Executive Committee. Ken has also spoken at numerous conferences, such as JavaOne, ApacheCon, and ConFoo, among others. He has written blogs at www.omnijava.com, and you are reading his first book. Ken can always be reached on Twitter, @omniprof, and Mastodon, @omniprof@mastodon.social.

About the reviewers

Massimiliano Dessì has more than 22 years of experience in programming, mainly in Java and Go. He currently works on the Sardinia island for Red Hat in the Kie Cloud Team. He was one of the founders of JUG Sardegna in 2002. He has also authored *Spring 2.5 Aspect-Oriented Programming*, published by Packt.

Robert Scholte is a Java Champion and a recognized expert in the Java community. He has been involved with the Apache Maven project since 2011 and served as the project lead between 2016 and 2022. Robert is highly regarded for his technical expertise, problem-solving skills, and passion for software development. He has also been recognized for his leadership and mentorship abilities, and he regularly provides training and guidance to other developers. Robert is also an active speaker at industry conferences and has been recognized for his passion and dedication to the Java community.

Table of Contents

Preface	xvii

Part 1: The Java Development Environment

1

Understanding Java Distributions — 3

Technical requirements	4
A little history	4
What makes Java special?	5
Why are there many distributions of Java?	6
Which Java should you get?	7
How is Java licensed?	7
Why are there so many versions of Java?	8
Installing Java	9

As an admin	9
As a non-admin	10
What is in the box?	12
Compiling and executing a Java program	14
Assembling and packaging a Java application	15
Documenting Java classes	15
REPL	16
Summary	16
Further reading	17

2

Code, Compile, and Execute — 19

Technical requirements	19
The first program	20
JShell – REPL in Java	22
The two-step compile and execute process – javac and java/javaw	27
Launch Single-File Source-Code Programs	29

For Windows, macOS, and Linux	29
For macOS and Linux – Shebang files	31
Integrated development environments	32
Eclipse Foundation – Eclipse	35
Apache NetBeans	36
Microsoft Visual Studio Code	37

JetBrains IntelliJ IDEA   38
Which IDE should you use?   39

Summary   39
Further reading   40

3

The Maven Build Tool   41

Technical requirements   42
Installing Maven   42
Windows   42
Linux   43
macOS   44

Overview of Maven functionality   44
Dependency management   44
Maven plugins   45
Maven project layout   45
Java source code packages   46

The pom.xml configuration file   50

jar – Java archive   53
war – web archive   53
ear – enterprise archive   53
pom – POM   53
The build section   56

Running Maven   60
Command-line Maven   60
Running Maven in an IDE   63

Summary   63
Further reading   63

Part 2: Language Fundamentals

4

Language Fundamentals – Data Types and Variables   67

Technical requirements   67
Primitive data types   68
Type safety   68
Literal values   71
Integers   71
Floating point   72
Boolean   73
char   74

A special case – String   75
Naming identifiers   76

Constants   78
Operators   78
Casting   80
Overflow and underflow   82
Wrapper classes   83
The math library   85

Summary   85
Further reading   86

5

Language Fundamentals – Classes   87

Technical requirements	87	The package specifier	90
Class fields	88	**Understanding classes**	**91**
Understanding access control	88	constructor and finalize methods	92
Packages	89	Revising the compound interest program	94
The public specifier	89	Class organization based on functionality	98
The private specifier	89	**Summary**	**106**
The protected specifier	90	**Further reading**	**106**

6

Methods, Interfaces, Records, and Their Relationships   107

Technical requirements	107	**Understanding inheritance**	**115**
Understanding methods	**108**	The superclass of all objects, the Object class	119
Access control designation	108	**Understanding the class interface**	**121**
Static or non-static designation and the this reference	109	Abstract class versus interface	124
Override permission – final	110	Sealed classes and interfaces	125
Override required – abstract	111	**Understanding the record class**	**126**
Return type	111	**Understanding polymorphism**	**127**
Method name	112	**Understanding composition in classes**	**130**
Parameter variables	112	Association	130
Annotations	112	Aggregation	132
Exception handling – throws	113	**Summary**	**133**
Thread setting	114		
Generic parameters	114		

7

Java Syntax and Exceptions   135

Technical requirements	135	Code blocks	136
Understanding coding structures	**136**	Statements	138
		Expressions	138

Operators   138
Iteration   140
Decision-making   142

Handling exceptions   **146**
The stack trace   149
Ending the program   149

The throw and throws statements   150
The finally block   150
Creating your own exception classes   151

Summary   **152**
Further reading   **153**

8

Arrays, Collections, Generics, Functions, and Streams   155

Technical requirements   **156**
**Understanding the array data
structure**   **156**
**Understanding the Collections
Framework**   **158**
**Using sequential implementations
and interfaces**   **158**
ArrayList   158
LinkedList   159
ArrayDeque   159
The Collection interface   160
How to declare a collection   160

**Understanding Generics in the
Collections Framework**   **161**
**Using sequential implementations
and interfaces with Generics**   **162**
**Understanding Collections
Framework map structures**   **166**
HashMap   166
LinkedHashMap   168
TreeMap   169

Understanding functions in Java   **170**
Using streams in collections   **172**
Summary   **173**
Further reading   **174**

9

Using Threads in Java   175

Technical requirements   **176**
Creating Java native OS threads   **176**
Extending the Thread class   176
Implementing the Runnable interface   179
Creating a thread pool with ExecutorService 180
Implementing the Callable interface   182

Managing threads   185
Daemon and non-daemon threads   187
Thread priority   187

**Preventing race and deadlock
conditions in threads**   **188**
Race condition   188

Deadlock condition   192
Creating new virtual threads   196

10

Implementing Software Design Principles and Patterns in Java   199

Technical requirements   199
SOLID software design principles   200
S – Separation of concerns/single responsibility200
O – Open/closed   201
L – Liskov substitution   203
I – Interface segregation   204
D – Dependency inversion   205

Summary   197
Further reading   198

Software design patterns   208
Singleton   208
Factory   211
Adapter   213
Observer   215
Summary   218
Further reading   219

11

Documentation and Logging   221

Technical requirements   221
Creating documentation   222
Comments   222
Javadocs   223
Using logging   228

java.util.logging   229
Log4j2   233
Summary   234
Further reading   234

12

BigDecimal and Unit Testing   235

Technical requirements   236
Using BigDecimal   236
What is JUnit 5?   241
Testing with JUnit 5   241

Performing parameterized testing   246
Summary   247
Further reading   248

Part 3: GUI and Web Coding in Java

13

Desktop Graphical User Interface Coding with Swing and JavaFX   251

Technical requirements	251	Application	262
A brief history of Java GUIs	252	PrimaryStage	262
Financial calculator program design	252	Pane	262
Internationalization – i18n	254	Scene	262
		CSS style sheets	263
Using the Swing GUI framework	256	JavaFX bean	264
JFrame	257	BigDecimalTextField	265
JPanel	257	Controls	266
Event handlers	258	Binding	266
Document filter	259	**Summary**	**267**
Pattern matching with regular expressions	259	**Further reading**	**267**
Controls and panels	260		
Using the JavaFX GUI framework	261		

14

Server-Side Coding with Jakarta   269

Technical requirements	270	How does a servlet access the query string in a request?	281
Understanding the role of the Java application server	270	How does a servlet remember my data?	284
GlassFish 7.0	272	**Configuring deployment with the web.xml file**	**285**
Configuring a web project with Maven	274	**Summary**	**286**
Changes to the pom.xml file	275	**Further reading**	**287**
Understanding what a servlet does and how it is coded	277		
What happens when a servlet is requested?	281		

15

Jakarta Faces Application   289

Technical requirements   290
Configuring a Faces application   290
Creating an object managed by
Context Dependency Injection and
validated with Bean Validation   293
FinanceBean   293
Scopes   294
Calculations   296

Using XHTML, Facelets, and
Expression Language for rendering
pages   297
Deploying a Faces web application   301

Understanding the life cycle of a
Faces page   301
Summary   302
Further reading   303

Part 4: Packaging Java Code

16

Deploying Java in Standalone Packages and Containers   307

Technical requirements   308
Exploring what modular Java is   308
Creating a custom JRE with jlink   310
Packaging with an installer using
jpackage   310
Using the Docker container system   312

Working with Docker images   312
Creating a Docker image   314
Publishing an image   315
Summary   316
Further reading   316

Index   317

Other Books You May Enjoy   328

Preface

In 1980, after being married for just a few months, I purchased my first computer. Why? Because at the age of 26, I thought it could be the best toy I could ever own. It was an Apple][+. Within days of its arrival, I became obsessed with it. In the late 70s and early 80s, numerous magazines were dedicated to the new realm of personal computers and I bought most of them. I'd come home from work when I was a photography technician at Dawson College and typed in any program I could find. Three years later, I left Dawson and struck out on my own programming professionally, and seven years after that, I became a college instructor in computer science at Dawson College.

I mention this story because computer programming changed my life. Every day I woke up, and to this day, I still wake up to face new challenges and problems that I need to solve. You are likely a programmer for the same reason.

Few, if any, developers spend their career only programming in a single language or on a single operating system. This is why I wrote this book. If you already code in one language, you already know the basics of almost every programming language. This is not a book on programming for beginners. It is a book for beginners to Java programming.

The four parts of this book present the skills you need to be familiar with to be a successful Java programmer. Java is a language that is experiencing rapid evolution. This book covers the recent enhancements to the language for both desktop and server-side programming.

In this book, I present source code that is just as important to read as the text of the book. All the source code is available on GitHub. For this book to have its full impact, I ask that you download all the examples. Run the examples and enhance them. Much of what I have learned has come from working with code samples. I describe what I do with these samples as conducting experiments. I encourage you to conduct your own experiments with the sample code.

While I do briefly discuss **Integrated Development Environments** (**IDEs**), all the examples can be edited with a simple text editor and then run either directly from the command line or with the Maven build tool. If you already have a favorite IDE or plan to use one, know that all the code built for Maven will load into any IDE without needing you to change anything.

Java has been around for 27 years at the time of writing. Whenever it has been declared out of date, too verbose, or too complex, the Java community has contributed new ideas, syntax, and libraries that have become part of the language. I could write for pages and pages on how Java has changed since its introduction in 1996. I won't, but what I do wish to impress on you is that Java is a language you are always learning. This book is just a starting point.

Who this book is for

This book has been written for developers with experience with languages other than Java. It is assumed that you have worked with one or more languages in the last few years. Recently you had either been assigned a new or existing Java project, yet that is not your background. Or maybe you just want to add to your resume. Think of this book as a beginner's book for experienced developers.

What this book covers

Chapter 1, Understanding Java Distributions, presents a little history of the language, followed by how Java is distributed, downloaded, and installed for Windows, Linux, and macOS.

Chapter 2, Code, Compile, and Execute, prepares you to code, compile, and execute your first program. An IDE is not required, and you just need a text editor, and the latest version of Java installed on your PC.

Chapter 3, The Maven Build Tool, shows you how to manage programs that can consist of many files along with dependencies on external libraries is the purview of the Maven build tool.

Chapter 4, Language Fundamentals – Data Types and Variables, provides an introduction to what data types are available in Java and the operations we can perform on them.

Chapter 5, Language Fundamentals – Classes, introduces object-oriented programming as carried out in Java and revolves around the structure called a class.

Chapter 6, Methods, Interfaces, Records, and Their Relationships, examines the structures in which Java code is written. You will learn that classes have an interface, the public members of the class. An interface class enforces an interface, while a record represents an immutable data class. The relationships between classes are described.

Chapter 7, Java Syntax and Exceptions, introduces you to the low-level syntax of Java employed in methods, and what happens when things go wrong, an exception is thrown.

Chapter 8, Arrays, Collections, Generics, Functions, and Streams, examines how we process multiple data elements in Java using arrays and collections. Generics enhance type safety, functions enhance the processing of elements, and streams provide an alternative to loops for processing multiple items.

Chapter 9, Using Threads in Java, introduces you to one of Java's greatest strengths, its inherent support of threads. The new threading approach called virtual threads is also examined.

Chapter 10, Implementing Software Design Principles and Patterns in Java, presents software design principles that provide guidance on how you construct your classes and how your objects should interact.

Chapter 11, Documentation and Logging, presents how you can document what a program does in the source code. Information about a program is shown while it is running using logging.

Chapter 12, BigDecimal and Unit Testing, demonstrates that when the accuracy of calculations, typically for currency, is required, then BigDecimal is the answer. Determining whether your program is delivering the correct results is the role of unit testing.

Chapter 13, Desktop Graphical User Interface Coding with Swing and JavaFX, presents the two Java **graphical user interface** (GUI) libraries, Swing and JavaFX, which support GUIs. You will see the same business logic presented using each library.

Chapter 14, Server-Side Coding with Jakarta, introduces you to backend web programming in Java. You will set up your application server and deploy Java code in the format of a servlet.

Chapter 15, Jakarta Faces Application, introduces you to server-side rendering of web pages. The application from *Chapter 13, Desktop Graphical User Interface Coding with Swing and JavaFX,* will now be presented as a Faces application.

Chapter 16, Deploying Java in Standalone Packages and Containers, shows we can distribute our applications. You will see how we can create installers for desktop programs and Docker containers for web applications commonly used for cloud deployment.

To get the most out of this book

There is just one assumption I have for you, and it is that you have a few years of experience working in any imperative language. If you are unsure what this means, then consider that JavaScript is an imperative language while HTML is a declarative language. Experience with an object-oriented, imperative language such as C# or C++ could be helpful but is not required.

At the time of this writing, Java 17 was the **long-term support** (LTS) version. All the code in this book will work in Java 17 and all subsequent versions. Versions prior to Java 17 may not work.

Software/hardware covered in the book	Operating system requirements
Java JDK 17 or greater	Windows, macOS, or Linux
Java JDK 19 or greater	Windows, macOS, or Linux
Maven 3.8.6 or greater	Windows, macOS, or Linux
GlassFish 7.0 Application Server	Windows, macOS, or Linux
Wix Toolset	Windows
Xcode command line tools	macOS
Docker Desktop	Windows, macOS, or Linux
The `rpm-build` package	Red Hat Linux
The `fakeroot` package	Ubuntu Linux
A text editor of your choice	Windows, macOS, or Linux
A web Browser of your choice	Windows, macOS, or Linux

If you are using the digital version of this book, we advise you to type the code yourself or access the code from the book's GitHub repository (a link is available in the next section). Doing so will help you avoid any potential errors related to the copying and pasting of code.

Download the example code files

You can download the example code files for this book from GitHub at `https://github.com/PacktPublishing/Transitioning-to-Java`. If there's an update to the code, it will be updated in the GitHub repository.

We also have other code bundles from our rich catalog of books and videos available at `https://github.com/PacktPublishing/`. Check them out!

Download the color images

We also provide a PDF file that has color images of the screenshots and diagrams used in this book. You can download it here: `https://packt.link/fyH3U`.

Conventions used

There are a number of text conventions used throughout this book.

`Code in text`: Indicates code words in text, database table names, folder names, filenames, file extensions, pathnames, dummy URLs, user input, and Twitter handles. Here is an example: "We begin with a class that implements `Runnable`. `actionCounter` is the number we will count down from in the thread."

A block of code is set as follows:

```
private final double principal = 100.0;
private final double annualInterestRate = 0.05;
private final double compoundPerTimeUnit = 12.0;
private final double time = 5.0;
```

When we wish to draw your attention to a particular part of a code block, the relevant lines or items are set in bold:

```
private final NumberFormat currencyFormat;
private final NumberFormat percentFormat;
```

Any command-line input or output is written as follows:

```
brew install openjdk@17
```

Bold: Indicates a new term, an important word, or words that you see onscreen. For instance, words in menus or dialog boxes appear in **bold**. Here is an example: "When you click on **Accept**, the code from the JShell editor will be transferred to JShell."

> **Tips or important notes**
> Appear like this.

Get in touch

Feedback from our readers is always welcome.

General feedback: If you have questions about any aspect of this book, email us at customercare@ packtpub.com and mention the book title in the subject of your message.

Errata: Although we have taken every care to ensure the accuracy of our content, mistakes do happen. If you have found a mistake in this book, we would be grateful if you would report this to us. Please visit www.packtpub.com/support/errata and fill in the form.

Piracy: If you come across any illegal copies of our works in any form on the internet, we would be grateful if you would provide us with the location address or website name. Please contact us at copyright@packt.com with a link to the material.

If you are interested in becoming an author: If there is a topic that you have expertise in and you are interested in either writing or contributing to a book, please visit authors.packtpub.com.

Share Your Thoughts

Once you've read *Transitioning to Java*, we'd love to hear your thoughts! Scan the QR code below to go straight to the Amazon review page for this book and share your feedback.

https://packt.link/r/1-804-61401-7

Your review is important to us and the tech community and will help us make sure we're delivering excellent quality content.

Download a free PDF copy of this book

Thanks for purchasing this book!

Do you like to read on the go but are unable to carry your print books everywhere? Is your eBook purchase not compatible with the device of your choice?

Don't worry, now with every Packt book you get a DRM-free PDF version of that book at no cost.

Read anywhere, any place, on any device. Search, copy, and paste code from your favorite technical books directly into your application.

The perks don't stop there, you can get exclusive access to discounts, newsletters, and great free content in your inbox daily

Follow these simple steps to get the benefits:

1. Scan the QR code or visit the link below

https://packt.link/free-ebook/9781804614013

2. Submit your proof of purchase
3. That's it! We'll send your free PDF and other benefits to your email directly

Part 1: The Java Development Environment

This part will introduce you to the development environment you will use throughout the book. While you may use an **integrated development environment** (**IDE**) tool, it is not necessary. As you will learn, all you will need is Java, a text editor, and the Maven build tool.

This part contains the following chapters:

- *Chapter 1, Understanding Java Distributions*
- *Chapter 2, Code, Compile, and Execute*
- *Chapter 3, The Maven Build Tool*

1

Understanding Java Distributions

In this chapter, we will examine how the **Java** language came about and how it is managed. While the word *Java* is used throughout this book, please note that I am referring to the **Java Standard Edition** or **Java SE**. There are numerous versions and distributions of Java, and this sometimes leads to confusion over which one to use. Is Java free, or do I have to license it? Can I include a Java runtime with my software? Can I distribute my own version of Java? These and other questions will be answered in this chapter.

You will learn how to install Java on Linux, macOS, and Windows. The significant tools that are part of the Java installation are highlighted in this chapter and will be used in later chapters.

We will cover the following topics in this chapter:

- A little history
- What makes Java special?
- Why are there many distributions of Java?
- Which Java should you get?
- How is Java licensed?
- Why are there so many versions of Java?
- Installing Java
- What's in the box?

Let's begin with a quick history lesson.

Technical requirements

To code in Java or run Java programs on your desktop, you need a computer and operating system that supports the **Java Development Kit (JDK)**. There are JDKs available for different operating systems and different **central processing units (CPUs)**. If you are running the Windows operating system, your only concern is whether you are running 32-bit or 64-bit. On macOS, there are versions of Java for both Intel and Apple (ARM) CPUs. If your operating system is Linux, there are more variations depending on your computer's hardware. There is even a version of Java for IBM mainframes that run Linux.

The only other hardware requirement is the amount of RAM on your system. I have run Java applications on a Raspberry Pi 3 Model B with just 1 GB of RAM. As a developer, you are doing more than just running programs. You run editors, compilers, web servers, database servers, and your usual software. Doing this needs memory. I recommend a minimum of 8 GB for a development system; 16 GB is ideal, and 32 GB might let you play games while you code.

A little history

Java did not start as a language called Java. In the early 1990s, the company **Sun Microsystems**, known for its SPARC workstations and the Solaris operating system, saw potential in the consumer electronics space. They put together a team of engineers to develop products in this space under the title **Green Project**. Their first device was called the **Star7**, a small handheld computer that used a custom version of Solaris. The Star7 is the first personal digital assistant, preceding the Apple Newton by a year. As part of the development of Star7, a language was created. James Gosling, a Canadian software engineer working for Sun, led a team that developed this new language for the Star7, called **Oak**. The Star7 never went into production, but Oak was destined to take over the world.

One of the consumer electronics targets Sun hoped that the Star7 could be used for was set-top boxes for the cable TV industry. They set up a company called FirstPerson and made a bid to develop a set-top box for the cable provider Time Warner. They lost the bid. While they were unsuccessful in bringing the Star7 to market, they saw potential in the Oak language. The only problem was that Oak was already trademarked.

There are numerous stories about how Oak became Java. Was it named after their favorite beverage or an island in Indonesia? Java was just 1 of 12 possible names. The names were turned over to the legal department for trademark searches. Of the names on the list given to the legal team, Java was the fourth name on the list and the first to pass the legal review. Java became the new name for Oak.

The early 1990s was also the time that the **World Wide Web (WWW)** became available to anyone with an internet connection. The Green team developed a browser called WebRunner coded with Java as a showcase for the language. This browser could run programs called Applets, which were written in Java. Java 1.0 was introduced to the world in 1995, and WebRunner was renamed HotJava. Netscape also licensed Java for their Navigator browser.

In 1998 Java 1.2, also referred to as Java 2, was introduced. Among many new features was the **Swing GUI library**, which significantly improved writing desktop GUI programs that ran independently from a browser. The **Java EE** platform was released in 1999 as **J2EE**. It was used to develop Java web servers. Now, you could write Java programs that responded to requests from a browser and were run on a web server. The rest, as the expression goes, is history.

What makes Java special?

Java was designed by Gosling and his team to address shortcomings they perceived in **C++**. The most significant of the shortcomings was memory management. In C++, variables of type pointer were used to allocate memory for objects. Once an object was no longer needed, the developer's responsibility was to release or deallocate the memory. Forgetting to do so resulted in memory leaks. A leak is a block of memory marked as *in use* but no longer accessible by a pointer. While Java still required you to allocate memory, you did not need to deallocate it. A process called the **garbage collector** tracked all memory allocations. When a pointer, named a reference in Java, went out of scope, the garbage collector would release its memory automatically. There are five garbage collectors available. The Parallel Garbage Collector is the default general-purpose collector. Serial Garbage Collector, CMS Garbage Collector, G1 Garbage Collector, and Z Garbage Collector use algorithms for specific types of applications such as those requiring low latency or requiring only a single thread.

However, garbage collection is not the most significant feature of Java. What sets Java apart from its predecessors, **C** and **C++**, is that Java programs do not execute directly in the computer's operating system. Instead, compiled Java programs, called **bytecode**, execute inside another process called the **Java virtual machine (JVM)**.

The JVM is a software simulation of a computer. The bytecode is the machine language of this simulated machine. The JVM then translates the bytecode into the machine language of the underlying computer.

The JVM is responsible for optimizing the code and performing garbage collection.

Native languages such as C and C++ are directly compiled into the machine language of the CPU coupled with the computer's operating system it will run on. Any libraries used must also have been compiled for a specific CPU and operating system. This means that a program compiled for an Intel CPU running Windows or an Apple M1 CPU running a specific version of macOS must be recompiled for an Intel CPU running Linux.

Java turns this concept on its head. Code that you write in Java and compile into bytecode can run on any hardware and operating system unchanged if there is a JVM for that computer. Java describes itself as a *Write Once Run Anywhere* language. This means that a Java application written on and for an Intel CPU will also run on an ARM-based system unchanged and without the need to recompile if there is a JVM for that platform.

In *Chapter 4, Language Fundamentals – Data Types and Variables*, and *Chapter 5, Language Fundamentals – Classes*, we will examine the syntax of the Java language.

Java is not the only language that runs in the JVM. More languages were developed to take advantage of the JVM while at the same time taking a different approach and syntax from Java. Here are four of the most widely used ones:

- Scala
- Kotlin
- Groovy
- Clojure

We now know what makes Java special in relation to languages that do not have a virtual machine. What can be confusing is that there is not just one version of Java distributed by just one company. Why? Let's take a look at that next.

Why are there many distributions of Java?

Java was first released as proprietary software. In 2006, Sun Microsystems created an open source version of Java called the **OpenJDK** with a GNU General Public License allowing developers to change and share the program. Sun (and later, the new owner, Oracle) retained Java-related intellectual property and copyrights.

One way to describe Java is to state that only JDKs and runtimes are considered Java if they pass an extensive suite of tests called the **Technology Compatibility Kit** (**TCK**). While Java was designated open source, initially, TCK was not. It needed to be licensed, for a fee, from Oracle. This resulted in very few companies making their own branded version of Java.

Today, however, it is possible to get access to the TCK without paying a fee. You have to make a formal request to Oracle, along with presenting several supporting documents explaining why you require access to the TCK. A screening committee will review your request and decide whether to grant you access to the TCK. At the time of writing, 27 organizations have signed the **OpenJDK Community TCK License Agreement** (**OCTLA**) and have access to the TCK.

So, why do companies still distribute their own branded version of Java? The simplest answer is to provide support to clients who wish to use Java in situations where the distributor may have more experience in a particular domain. Cloud providers such as Microsoft and Amazon have their own branded versions of Java that have been optimized for their cloud infrastructure. BellSoft, the distributor of the Liberica distribution, is one of the leaders involved in ARM versions of Java. While it might not make much of a difference which distribution you choose, the distribution your clients will use is significant.

Regardless of the distributor, the language is maintained by Oracle. A well-established process allows anyone to propose changes to the language. Through the **Java Community Process** (**JCP**), all changes, additions, and removals from the language are carefully reviewed.

The actual coding of changes to the JDK is primarily the responsibility of developers working for Oracle. Consider joining the JCP to keep abreast of changes and contribute to the language.

Let's move on and look at which version you should use since you do not have any experience with the language.

Which Java should you get?

All distributions of Java since Java 11, including Oracle's distribution, are based on the OpenJDK source code. It should not matter whose distribution of Java you choose if it has passed the TCK. If you do not have a distribution you must use, then I recommend the Eclipse Adoptium version called **Temurin**. This version has passed the TCK. Java is a registered trademark, so the word cannot be used for distributions other than from Oracle, hence the name Temurin. If you are curious about where this name came from, I will give you a hint – it is an anagram.

You might think that the obvious choice for a Java distribution would be an Oracle-branded version. This was pretty much the case until the final release of Java 8. With this release, Oracle required companies that distributed Java as part of their commercial offerings to purchase a commercial support license for access to updates to Java. Starting with Java 11, Oracle required commercial licensees to purchase a subscription for every developer. Personal use of the Oracle-branded JDK has remained free to use, though.

This gets confusing because should you choose to use the OpenJDK distribution or any other distribution based on the OpenJDK except for Oracle's, there are no fees required for commercial distribution. With the release of Java 17, Oracle changed its licensing again. Now called the Oracle **No-Fee Terms and Conditions** (**NFTC**), this now allows you to use Oracle's Java for the development of software and then distribute this version of Java with your program without the need for a subscription or fee. This is only applicable to versions of Java starting at 17. Versions from 8 to 16 are still subject to the licenses.

How is Java licensed?

If you plan to use Java to develop software commercially, then how it is licensed is important to you. As already stated, the OpenJDK carries the GNU General Public License version 2, commonly referred to as the **GPLv2**. The GPL is widely used in open source software. At its most basic level, it requires any software that uses GPL licensed code to also be subject to the GPL. This means that any software you create must make its source code available under the same conditions. Copyright and intellectual property rights stay with the author of the work, either Oracle or you.

Java's GPLv2 carries with it the Classpath Exception, also called the linking exception. A classpath, like an operating system path, is the location of classes and packages that the JVM and Java compiler will use. Under this exception, you do not need to supply the source code when you distribute your application. The software you write that is linked to Java does not require a GPLv2 license. It can remain proprietary and cannot be freely used like GPL software. You choose the licensing for the code that you generate.

Why are there so many versions of Java?

Java is constantly evolving – bug fixes, enhancements, and new features are in continuous development. Java was initially numbered as 1 plus a version number. The first nine versions starting in 1996 and until 2014 were 1.0, 1.1, 1.2, 1.3, 1.4, 1.5, 1.6, 1.7, and 1.8. Between each of these versions, there was a third number that represented an update rather than a major revision, such as 1.8_202.

Starting with Java 1.8, then subsequently named Java 8, here is the timeline of Java versions:

Version	Distribution date	Version	Distribution date
Java SE 8 LTS	March 2014	Java SE 18	March 2022
Java SE 9	September 2017	Java SE 19	September 2022
Java SE 10	March 2018	Java SE 20	March 2023
Java SE 11 LTS	September 2018	Java SE 21 LTS	September 2023
Java SE 12	March 2019	Java SE 22	March 2024
Java SE 13	September 2019	Java SE 23	September 2024
Java SE 14	March 2020	Java SE 24	March 2025
Java SE 15	September 2020	Java SE 25 LTS	September 2025
Java SE 16	March 2021	Java SE 26	March 2026
Java SE 17 LTS	September 2021	Java SE 27	September 2026

Table 1.1 – Timeline of Java versions

You will see several versions designated **LTS**, short for **Long Term Support**, by Oracle. These versions are expected to be supported with bug fixes and security updates for at least 8 years. The non-LTS versions, also called **feature releases**, are accumulative fixes, updates, and preview features. Support for these versions is expected to last only until the next non-LTS or LTS version is released. Companies that have their own Java distribution may provide support for longer than Oracle does.

LTS versions are typically what many organizations prefer to use for their products. Java 8, released in March 2014, is still supported and will be until December 2030. Subsequent LTS versions are being supported for just 8 years but, as already mentioned, may be supported by other Java distributors for a longer period. The current schedule for new releases of Java has an LTS version every 2 years. A non-LTS version is released every 6 months.

If you plan to develop server-side software, you must use an LTS version. Libraries required for server-side are written to use a specific LTS version. When a new LTS version is released, it might take some time for all such libraries to be updated as is currently the case with LTS Java 17. As I write this, most server-side applications are running Java 11 and some still even use Java 8.

What has contributed to Java's success is evident in the continuing and now regular cadence of releases. This ensures that Java continues to be a state-of-the-art language.

Installing Java

Installing Java is a simple process. As a developer, you will install the JDK from any of the distributors. Most Java distributors have packaged Java with an installer and as a compressed file without an installer that you can download. The choice depends on your OS, CPU, and whether you are the administrator or superuser and can use an installer. Or, you are a client and can only install the compressed file.

With your distribution and version decided, you are ready to install Java as an admin and non-admin.

As an admin

As an admin, you can install Java for all users of the computer in the following ways.

Windows

Download the appropriate (32- or 64-bit) .msi file for Java from https://adoptium.net/. This type of file contains an installer that will place Java in the folder of your choice and configure the appropriate environment variables. Just double-click on the .msi file after it is downloaded. The Windows installer will lead you through the installation.

macOS

You have two options for installing Java for macOS. The first is to download the .pkg file for Mac that includes an installer. Just double-click on the .pkg file after it is downloaded. The Apple installer will lead you through the installation.

The second is to use **HomeBrew**, a command-line utility for managing new software and updates, which will download and install Java.

With HomeBrew installed, you can install the OpenJDK version with the following:

```
brew install openjdk@17
```

To install the Eclipse Temurin version of Java 17, use the following:

```
brew tap homebrew/cask-versions
brew install --cask temurin17
```

Linux

On Linux, you use the apt install command-line tool. You must be a superuser/admin to use this tool. You also include the distribution and version you require. You install the OpenJDK Java at the command line with the following:

```
sudo apt install openjdk-17-jdk
```

To install the Eclipse Temurin version of Java, use the following:

```
sudo apt install temurin-17-jdk
```

Verifying installation

Once the installation is complete, verify that Java works by issuing the following command:

```
java -version
```

If it shows you the version and distribution name of Java that you just installed, you are done and ready to code Java. The version number may be different depending on when you download Java or use apt install. Here is what you should see:

Windows

```
>java -version
openjdk version "17.0.3" 2022-04-19
OpenJDK Runtime Environment Temurin-17.0.3+7 (build 17.0.3+7)
OpenJDK 64-Bit Server VM Temurin-17.0.3+7 (build 17.0.3+7,
mixed mode,
    sharing)
```

Linux and macOS

```
$ java -version
openjdk version "17.0.3" 2022-04-19
OpenJDK Runtime Environment Temurin-17.0.3+7 (build 17.0.3+7)
OpenJDK 64-Bit Server VM Temurin-17.0.3+7 (build 17.0.3+7,
mixed mode,
    sharing)
```

If it tells you that it cannot find Java, then follow the instructions given in the coming *Configuring environment variables* section for setting up the environment variables.

As a non-admin

If you are not an admin, then you can still install Java but only you will be able to use it.

Windows

Windows users can download the appropriate .zip file version and unzip it in the desired folder.

Linux and macOS

Download the appropriate .tar.gz file version for either Linux or macOS. Once downloaded, use the following command line. The only difference between Linux and macOS is the name of the file.

For Linux, use the following:

```
tar xzf OpenJDK17U-jdk_x64_linux_hotspot_17.0.3_7.tar.gz
```

For macOS, use the following:

```
tar xzf OpenJDK17U-jdk_x64_mac_hotspot_17.0.3_7.tar.gz
```

Configuring environment variables

There are two environment variables that need to be set. While the environment variables are the same on Windows, Linux, and macOS, the process of setting them differs.

The first environment variable is JAVA_HOME. Certain Java processes, such as web servers, need to know where Java is installed to be able to access specific components in the JDK. It must be assigned the full path to the folder in which you have installed Java.

The second environment variable is PATH. When a program is run from the command line, the OS will look for an executable file in the current directory. If it is not found, then it will go through every directory in the path to look for it.

You will have to enter these commands every time you open a console. Adjust the command based on your login name and the version of Java you are installing. While you can install multiple versions of Java, only one can be used for JAVA_HOME and PATH:

Windows

```
set JAVA_HOME= C:\devapp\jdk-17.0.2+8
set PATH=%JAVA_HOME%\bin;%PATH%
```

Adjust the path to the folder created when you unzipped the Java file. You can also place these two lines in a batch file that you can run every time you open a console to code in Java:

Linux

```
export JAVA_HOME=/home/javadev/java/jdk-17.0.2+8
export PATH=$JAVA_HOME/bin:$PATH
```

This assumes that you are logged in as javadev and you are placing Java in a directory called java. These two lines can be added to your .profile file in your home directory so that they execute every time you log in.

macOS

```
export JAVA_HOME=/Users/javadev/java/jdk-17.03+7/Contents/Home
export PATH=$JAVA_HOME/bin:$PATH
```

This assumes that you are logged in as javadev and you are placing Java in a directory called java. These two lines can be added to your .bash.profile file in your home directory so that they execute every time you log in.

Verifying installation

You can quickly determine whether your installation of Java is correct. Open a command or console window on whatever system you are using. If you performed the non-admin installation, then ensure that JAVA_HOME and PATH have been updated and set. In the command window, enter the following:

```
java -version
```

If the installation was successful, the output, if you installed the OpenJDK, will be as follows:

```
openjdk version "17.0.3" 2022-04-19
OpenJDK Runtime Environment (build 17.0.3+7-Ubuntu-
0ubuntu0.20.04.1)
OpenJDK 64-Bit Server VM (build 17.0.3+7-Ubuntu-
0ubuntu0.20.04.1, mixed mode, sharing)
```

The output, if you installed the Temurin JDK, will be as follows:

```
openjdk version "17.0.3" 2022-04-19
OpenJDK Runtime Environment Temurin-17.0.3+7 (build 17.0.3+7)
OpenJDK 64-Bit Server VM Temurin-17.0.3+7 (build 17.0.3+7,
mixed mode,
    sharing)
```

Your installation is now complete and verified. Let's now examine some of the files that you just installed.

What is in the box?

The JDK contains the programs and libraries necessary to compile your source code into bytecode and then execute your code in the JVM program. It also includes numerous tools that support your work as a developer.

There is a second packaging of Java, called the **Java Runtime Edition** (JRE). This smaller package only contains the components necessary to run Java bytecode and not the Java compiler.

Java 9 introduced a new way to package Java applications that rendered the JRE superfluous. As of Java 11, Oracle no longer distributes a JRE for both their distribution and the OpenJDK. Certain distributions from other companies may still provide a JRE for the current versions of Java. We will look at the modular approach to packaging Java applications in a later chapter.

The installation of Java will take up approximately 300 MB of disk space depending on the underlying OS.

The following are the directory structures from a Linux and Windows Java installation. The first is for Ubuntu but will be almost identical on all Linux and macOS installations.

The directory structure for Ubuntu 20.04.4 LTS is as follows:

```
$ ls -g -G
total 36
-rw-r--r--  1 2439 Apr 19 17:34 NOTICE
drwxr-xr-x  2 4096 Apr 19 17:34 bin
drwxr-xr-x  5 4096 Apr 19 17:33 conf
drwxr-xr-x  3 4096 Apr 19 17:33 include
drwxr-xr-x  2 4096 Apr 19 17:33 jmods
drwxr-xr-x 72 4096 Apr 19 17:33 legal
drwxr-xr-x  5 4096 Apr 19 17:34 lib
drwxr-xr-x  3 4096 Apr 19 17:33 man
-rw-r--r--  1 1555 Apr 19 17:34 release
```

The directory structure for Windows Enterprise 11 Version 21H2 is as follows:

```
>dir

2022-03-29  11:28 AM    <DIR>          .
2022-05-03  05:41 PM    <DIR>          ..
2022-03-29  11:28 AM    <DIR>          bin
2022-03-29  11:28 AM    <DIR>          conf
2022-03-29  11:28 AM    <DIR>          include
2022-03-29  11:28 AM    <DIR>          jmods
2022-03-29  11:28 AM    <DIR>          legal
2022-03-29  11:28 AM    <DIR>          lib
2022-03-29  11:28 AM            2,401 NOTICE
2022-03-29  11:28 AM            1,593 release
```

If we investigate the bin folder, we will find several executable programs that Java refers to as its tools. On a Windows system, they all have the .exe extension; on Linux and macOS, they appear as names only. In this chapter, we will discuss the following tools:

- jar
- java
- javadoc
- jlink
- jmod
- jpackage
- jshell
- javaw

These are tools we will be using in the coming chapters. See the tool specification link in the *Further reading* section for details on all the tools included in the JDK.

We have divided these tools into the following categories:

- Compiling and executing a Java program
- Assembling and packaging a Java application
- Documenting Java classes
- **Read, Evaluate, Print, and Loop (REPL)**

Let's look at each of these categories.

Compiling and executing a Java program

These are the tools that take us from source code to running a Java program. Some important tools are as follows.

javac

This is the Java compiler. Its role is to compile a Java source code file that ends in .java into a bytecode file that ends in .class.

java or javaw.exe

This is the tool that starts up the JVM process and then executes the bytecode file in the process. When using java, a console window will open and remain open until the JVM process ends. The javaw tool also starts up the JVM and executes a bytecode program. It will not open a console.

Windows users typically do not expect a console to open as they may have never seen one or interacted with one. If you wish to create a Windows shortcut to run a Java program, you will use `javaw.exe program.class`.

We will examine these three commands in *Chapter 2, Code, Compile, and Execute*.

Assembling and packaging a Java application

A Java program can be constructed of hundreds, thousands, or even more `.class` files. In these cases, it is necessary to assemble all these files along with any supporting files (such as images) into a single file. Some of the tools that do so are the following.

jar

A Java program or library usually consists of multiple `.class` files. To simplify the delivery of such programs, the `jar` tool combines all the class files of an application or library into a single file that uses ZIP compression and has the `.jar` extension. A `.jar` file can be designated as executable. In this case, you already have a Java JDK or JRE installed. We will see how this tool is used in *Chapter 2, Code, Compile, and Execute*.

jmod and jlink

Java 9 introduced the concept of modular Java, an approach to assembling Java applications that uses a new format for combining class files called `.jmod` files. These files are like `.jar` files in that they are ZIP compressed files. The `jmod` tool creates `.jmod` files.

Until Java 9 came along, a single file called `rt.jar` held all the Java libraries. Starting with Java 9, the Java libraries exist as individual `.jmod` files. Java packages the JVM file as a `.jmod` file. What this means to a developer is that it is possible to distribute Java applications that include the JVM and only those components of Java that must be available to execute your program. It is now unnecessary to have the JDK or JRE pre-installed as all that you need to execute the program is in the archive. You still need the `jar` tool to construct such an executable as you cannot execute `.jmod` files. We will see how these two tools are used in *Chapter 16, Deploying Java in Standalone Packages and Containers*.

jpackage

The `jpackage` tool creates native applications that hold a Java application and a Java runtime. It is used with either `.jar` or `.jmod` files. The output is an executable file, such as `.msi` or `.exe` for Windows, or a `.dmg` file for a macOS system. We will see how this tool is used in *Chapter 16, Deploying Java in Standalone Packages and Containers*.

Documenting Java classes

Having carefully documented your code, Java has a tool for gathering all your comments for easy access for other developers who may use your code.

javadoc

Documenting code has always been an issue in almost every language. Java takes a unique approach to encouraging documentation. If you comment your code in a specific format (which we will examine in *Chapter 4, Language Fundamentals – Data Types and Variables*), the `javadoc` tool will generate an HTML page for every class that you create. On this page, you will find all public members of the class presented.

Look at the `javadoc` page for the `ArrayList` class at `https://docs.oracle.com/en/java/javase/17/docs/api/java.base/java/util/ArrayList.html`. Everything you see on this web page was written into the source code file that was then converted to the HTML page you are looking at. We will examine this tool in *Chapter 11, Documentation and Logging*.

REPL

A REPL tool is one that supports the execution of code, one line at a time.

jshell

The `jshell` tool allows you to write and execute individual Java statements without the need for the usual decorations of classes and methods. This can be quite useful for learning Java. It can execute code line by line as you write it. We will examine `jshell` in *Chapter 2, Code, Compile, and Execute*.

Summary

In this chapter, we have learned a little about Java's history, how it is licensed, and why there are so many distributions and versions of Java. You now understand Java as a development tool and know how to select a Java distribution and version. We saw how we could install Java on our computers regardless of the OS. In *Chapter 12, BigDecimal and Unit Testing*, we will also examine how to install Java in a Docker container. We wrapped up the chapter with a look at nine Java tools that come with the JDK; we will see them again in later chapters. We will learn more about these tools in those chapters.

In *Chapter 2, Code, Compile, and Execute*, we will learn how we write, compile, link, and execute Java programs. Coding with a plain text editor, `jshell`, and with an **integrated development environment** (**IDE**) will be our focus.

Further reading

- **How Java got its name:** `https://www.quora.com/How-Java-got-its-name/answer/Ashok-Kumar-1682`

- **GNU General Public License, version 2, with the Classpath Exception:** `https://openjdk.java.net/legal/gplv2+ce.html`

- **Oracle No-Fee Terms and Conditions (NFTC):** `https://www.oracle.com/downloads/licenses/no-fee-license.html`

- **Java® Development Kit Version 17 Tool Specifications:** `https://docs.oracle.com/en/java/javase/17/docs/specs/man/index.html`

- **JVM Garbage Collectors:** `https://www.baeldung.com/jvm-garbage-collectors`

2

Code, Compile, and Execute

With Java installed, we are almost ready to look at coding. Before we get to that, though, we need to learn how to code, compile, and execute Java applications. While an **integrated development environment** (**IDE**) will likely be what you will use for most of your work, understanding how to code without the hand-holding of an IDE is what makes the difference between a Java tinkerer and a Java professional.

In this chapter, we will look at working from the command line and then from some of the most widely used IDEs. This chapter will not be a tutorial on IDEs but rather a review of what they offer to a programmer. The fundamental operation of any IDE is very similar to that of the most commonly used IDEs. Before we examine the various ways to use Java, we will look at a small program that we will use. This book is a *Hello World!* free zone, which means that example number one does something useful.

The goal of this chapter is to make you familiar with the four approaches to compiling and executing Java code as well as introduce you to the IDE tools available to developers. We will be covering the following topics:

- The first program
- JShell – REPL in Java
- The two-step compile and execute process – `javac` and `java`/`javaw`
- Launch Single-File Source-Code Programs
- Integrated development environments

Technical requirements

To follow the examples in this chapter, you will need the following:

- Java 17 installed
- A text editor, such as Notepad

You can find the code files for this chapter on GitHub at `https://github.com/PacktPublishing/Transitioning-to-Java/tree/chapter02`.

The first program

Before we can learn how to compile and execute Java code, we need a Java program to work with. Our first program will calculate compound interest. There is a quote attributed to Albert Einstein, who is said to have stated, "*Compound interest is the eighth wonder of the world.*" Whether he ever said this remains in doubt. Regardless, calculating interest on interest is one of the most important financial calculations that can be performed. Here is the formula that we will implement:

$$A = P(1 + \frac{r}{n})^{nt}$$

Here, P is the principal amount deposited into a compound interest account, r is the interest rate typically expressed as an annual rate, n is the number of compounding periods (if compounded monthly, then the value is 12), and t is the time the money will compound for. This is expressed in years and must be divided by the number of compounding periods, which, in this case, will also be 12.

In *Part 2* of this book, we will examine the syntax and structure of the language. There, we will explore the code of this program. For this reason, we will just use this program, which is simple to understand. You can download it from this book's GitHub repository.

Here is the code for a simple program in Java that will calculate what a fixed amount of money will be worth after a length of time has passed at a fixed rate of interest:

```java
import java.text.NumberFormat;

public class CompoundInterest01 {

    private final double principal = 100.0;
    private final double annualInterestRate = 0.05;
    private final double compoundPerTimeUnit = 12.0;
    private final double time = 5.0; //

    private final NumberFormat currencyFormat;
    private final NumberFormat percentFormat;

    public CompoundInterest01() {
        currencyFormat =
          NumberFormat.getCurrencyInstance();
```

```java
        percentFormat = NumberFormat.getPercentInstance();
        percentFormat.setMinimumFractionDigits(0);
        percentFormat.setMaximumFractionDigits(5);
    }

    public void perform() {
        var result = calculateCompoundInterest();
        System.out.printf("If you deposit %s in a savings
                account " + "that pays %s annual interest
                compounded monthly%n" + "you will have
                after %1.0f years %s",
                currencyFormat.format(principal),
                percentFormat.format(annualInterestRate),
                time, currencyFormat.format(result));
    }

    private double calculateCompoundInterest() {
        var result = principal * Math.pow(1 +
            annualInterestRate / compoundPerTimeUnit,
            time * compoundPerTimeUnit);
        return result;
    }

    public static void main(String[] args) {
        var banker = new CompoundInterest01();
        banker.perform();
    }
}
```

All Java programs must consist of at least one structure known as a **class**. Whereas C++ allows you to mix the structured style and the object-oriented style, Java requires the latter style.

With that, you've seen your first complete Java program. If you are coming from a C++ or C# background, you likely understand how it works. If you don't have an **object-oriented programming (OOP)** background, you can look at it as a structured program. In *Part 2* of this book, we will explore the syntax of Java. Next, we will run this program in three different ways from the command line.

JShell – REPL in Java

Read-Eval-Print Loop (**REPL**) is an environment where code can execute one line at a time. REPL became a standard part of Java in version 9. It is implemented in a tool called JShell. It serves two purposes:

- It provides an environment for learning Java without any background in programming.
- It provides a way to quickly test concepts, syntax, and libraries.

As such, you can execute Java code without the need for the usual decorations.

There are two ways we can use JShell. The first is to just enter the code that's necessary to use the formula. Imagine that you want to verify the formula for compound interest, as shown in the source code. You can do that by just entering the necessary code to perform the calculation.

In the following code block, we have entered the four variable declarations with the values required for the calculation, followed by the line of code that performs the calculation and assigns it to a variable:

```
private final double principal = 100.0;
private final double annualInterestRate = 0.05;
private final double compoundPerTimeUnit = 12.0;
private final double time = 5.0; //
var result = principal * Math.pow(1 + annualInterestRate /
compoundPerTimeUnit, time * compoundPerTimeUnit);
```

Follow these steps in JShell to enter these five lines of code:

1. On Windows, open the Command Prompt. If you are working on a macOS/Linux system, go to the Terminal. If necessary, set the `Path` and `JAVA_HOME` values. Then, enter the `jshell` command. The console will look like this:

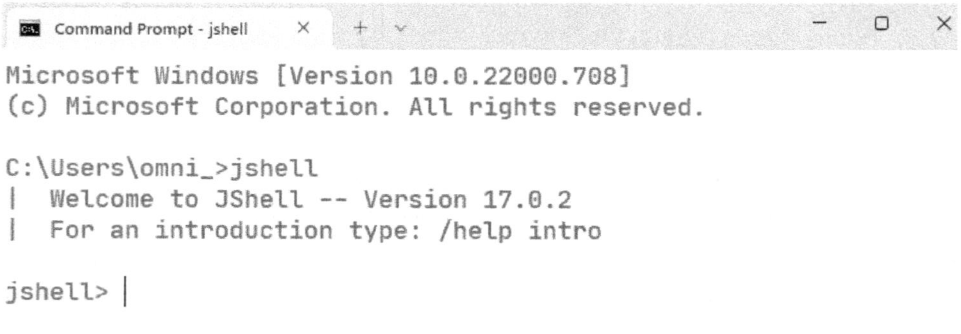

Figure 2.1 – Running JShell

2. Now, we can enter the following five lines of code:

```
Command Prompt - jshell        ×      + ∨                                              —    □    ×

Microsoft Windows [Version 10.0.22000.708]
(c) Microsoft Corporation. All rights reserved.

C:\Users\omni_>jshell
|  Welcome to JShell -- Version 17.0.2
|  For an introduction type: /help intro

jshell> private final double principal = 100.0;
principal ==> 100.0

jshell> private final double annualInterestRate = 0.05;
annualInterestRate ==> 0.05

jshell> private final double compoundPerTimeUnit = 12.0;
compoundPerTimeUnit ==> 12.0

jshell> private final double time = 5.0;
time ==> 5.0

jshell> var result = principal * Math.pow(1 + annualInterestRate / compoundPerTi
meUnit, time * compoundPerTimeUnit);
result ==> 128.33586785035118

jshell> |
```

Figure 2.2 – Executing one line at a time in JShell

Note that each line is being executed as it is being entered and JShell reports the value that's been assigned to each variable. Java does not format values on its own, so the result is a raw floating-point number.

Can we assume that since it was executed without any errors, the result is correct? Absolutely not! Especially for calculations, you need a second source for the result. This is where a spreadsheet is invaluable. Here is the result of the calculation after using the formula in Microsoft Excel:

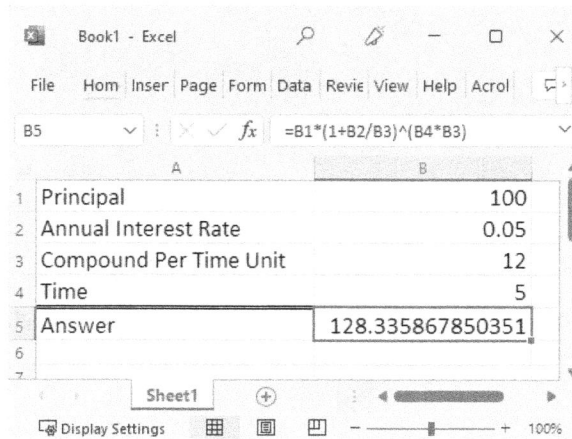

Figure 2.3 – The compound interest calculation in Excel

If any errors occurred while you entered your code, which means that your code doesn't match the result in the spreadsheet, then you can clear everything you entered into JShell with the `/reset` command and start over. JShell always preserves the last piece of code you entered unless you reset it. JShell also maintains a history of everything you entered and is not lost when you use `/reset`. **History**, which can be accessed with the up and down arrows, will just allow you to enter the code you used previously into what you are currently working on.

The second way that we can use JShell is by using an editor and making an entire program available to the tool. Before we provide JShell with the complete program, we must use the `/reset` command to remove what we entered previously.

To execute the entire program, follow these steps:

1. JShell has a basic editor that supports multi-line Java code. Use the `/edit` command to open the editor.

2. This editor cannot open files, so you will need to open the `CompoundInterest01.java` file in your text editor, copy the file's contents, and paste it into the JShell editor.

3. When you click on **Accept**, the code from the JShell editor will be transferred to JShell:

```
JShell Edit Pad                                      —    □    ×

import java.text.NumberFormat;

public class CompoundInterest01 {

    private final double principal = 100.0;
    private final double annualInterestRate = 0.05;
    private final double compoundPerTimeUnit = 12.0;
    private final double time = 5.0; //

    private final NumberFormat currencyFormat;
    private final NumberFormat percentFormat;

    public CompoundInterest01() {
        currencyFormat = NumberFormat.getCurrencyInstance();
        percentFormat = NumberFormat.getPercentInstance();
        percentFormat.setMinimumFractionDigits(0);
        percentFormat.setMaximumFractionDigits(5);
    }

    public void perform() {
        var result = calculateCompoundInterest();
        System.out.printf("If you deposit %s in a savings account "
                + "that pays %s annual interest compounded monthly%n"
                + "you will have after %1.0f years %s",
                currencyFormat.format(principal),
                percentFormat.format(annualInterestRate),
                time, currencyFormat.format(result));
    }

    private double calculateCompoundInterest() {
        var result = principal * Math.pow(1 + annualInterestRate /
                compoundPerTimeUnit, time * compoundPerTimeUnit);
        return result;
    }

    public static void main(String[] args) {
        var banker = new CompoundInterest01();
        banker.perform();
    }
}

              Cancel       Accept       Exit
```

Figure 2.4 – The program after being pasted into the default JShell editor

4. Now, you must click **Exit** to leave the editor, which will restore the JShell prompt.

5. Now, you can run the `main` method by entering `CompoundInterest01.main(null)` in the JShell prompt, which will cause the program to execute. The `main` method expects a parameter. Should you not have a parameter to pass to the `main` method, then you will automatically pass null:

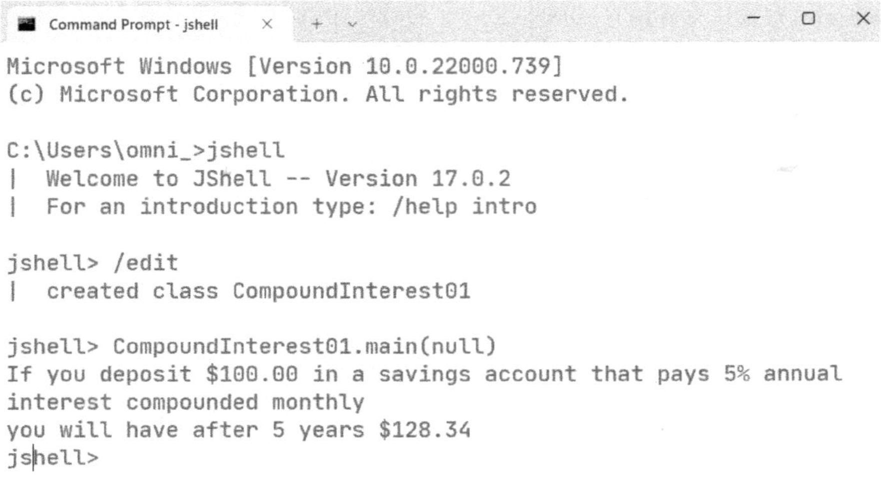

```
Command Prompt - jshell        ×    + ∨                          —   □   ×

Microsoft Windows [Version 10.0.22000.739]
(c) Microsoft Corporation. All rights reserved.

C:\Users\omni_>jshell
|  Welcome to JShell -- Version 17.0.2
|  For an introduction type: /help intro

jshell> /edit
|  created class CompoundInterest01

jshell> CompoundInterest01.main(null)
If you deposit $100.00 in a savings account that pays 5% annual
interest compounded monthly
you will have after 5 years $128.34
jshell>
```

Figure 2.5 – Running the program in JShell after it has been transferred from the editor

You do not have to use the basic editor in **JShell**. Here, you can use the `/set editor` command and include the path to the editor of your choice, as shown in the following screenshot:

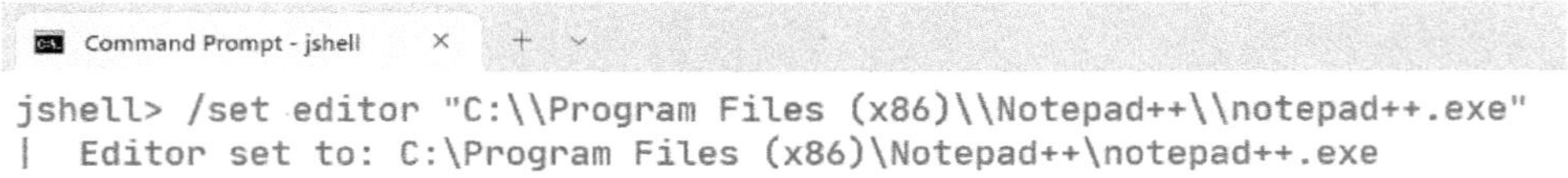

```
Command Prompt - jshell     ×    + ∨

jshell> /set editor "C:\\Program Files (x86)\\Notepad++\\notepad++.exe"
|  Editor set to: C:\Program Files (x86)\Notepad++\notepad++.exe
```

Figure 2.6 – Changing the editor in JShell

Here, I have set the editor to Notepad++ for Windows. Note that there are double backslashes for the path separators in the `/set editor` command as it is being entered on a Windows system. On a Linux or Mac system, the path separator is a forward slash and is not doubled. As there are spaces in the path, they must be enclosed in quotation marks.

When using an external editor with JShell, you must exit the editor to transfer what you have typed to JShell. This is because there is no *Accept* button.

The JShell tool can be useful for testing or learning a new feature or syntax in Java. It can also be quite useful for teaching Java to absolute beginners. Familiarize yourself with this tool as soon as possible.

The two-step compile and execute process – javac and java/javaw

The most common approach to running a Java program involves two steps. First, you must compile the code with javac and then execute the code in the **Java Virtual Machine** (**JVM**) with Java or on Windows with javaw.

The first step of preparing Java source code for execution is to compile it into **bytecode**. This is the machine language of the JVM. Every Java source file that is part of an application must be compiled into bytecode.

The second step is to execute the bytecode in a JVM. Unlike C or C++, there is no link step. The link step combines all compiled code into a single executable file. In Java, all the bytecode files must be on the classpath, which is the path to all bytecode files, and not necessarily combined into a single file. This may seem confusing as there is a tool called **jlink**, but its purpose is to combine the Java runtime with your code so that the end user does not need the Java version that was previously installed on their computer. We will examine **jlink** in *Chapter 16, Deploying Java in Standalone Packages and Containers*.

Let us compile and execute our CompoundInterest01 program:

1. First, place the CompoundInterest01.java file in a folder of its own. This is not a requirement, but it is easier to manage the code if it has its own folder.

2. Now, open a console in that folder; if you are not using an admin installation of Java, set the Path and JAVA_HOME properties as shown in the previous chapter. Now, you can compile the program with javac CompoundInterest01.java:

```
Command Prompt          ×   + ∨                              –   □   ×
Microsoft Windows [Version 10.0.22000.739]
(c) Microsoft Corporation. All rights reserved.

C:\PacktJavaCode\CompoundInterest01>javac CompoundInterest01.java

C:\PacktJavaCode\CompoundInterest01>dir
 Volume in drive C has no label.
 Volume Serial Number is 46B1-7989

 Directory of C:\PacktJavaCode\CompoundInterest01

2022-06-15  03:03 PM    <DIR>          .
2022-06-15  02:57 PM    <DIR>          ..
2022-06-15  04:02 PM             1,636 CompoundInterest01.class
2022-06-15  04:01 PM             1,433 CompoundInterest01.java
               2 File(s)          3,069 bytes
               2 Dir(s)  672,338,796,544 bytes free

C:\PacktJavaCode\CompoundInterest01>
```

Figure 2.7 – Compiling the Java program

If compiling your code does not result in any errors, then the compiler will not display anything in the console. The folder will now contain a new file called CompoundInterest01.class. Notice that it is larger than the Java source code file.

A Java class file contains the necessary **bytecode** and source code. The presence of the source code in this file supports the concepts of **reflection** and **introspection**.

Introspection allows you to write code that can examine the type or properties of an object at runtime, whereas reflection allows you to write code that can examine and modify the structure of an object at runtime. These two features are rarely used in business programming. .

3. With the code compiled, we can run it. Here, we are using the java executable to run our program.

    A common myth is to believe that the java executable is the JVM. It is not. It is the loader of the JVM, made up of other files that are part of the installation of the JDK. The JVM utilizes its ClassLoader component to locate and execute a bytecode file.

    Let us run the CompoundInterest01 program. In the console, enter java CompountInterest01. There's no need to include the .class extension as that is the only acceptable extension:

```
Command Prompt        ×    +  ˅                           –   □   ×

Microsoft Windows [Version 10.0.22000.739]
(c) Microsoft Corporation. All rights reserved.

C:\PacktJavaCode\CompoundInterest01>javac CompoundInterest01.java

C:\PacktJavaCode\CompoundInterest01>dir
 Volume in drive C has no label.
 Volume Serial Number is 46B1-7989

 Directory of C:\PacktJavaCode\CompoundInterest01

2022-06-15  03:03 PM    <DIR>          .
2022-06-15  02:57 PM    <DIR>          ..
2022-06-15  04:02 PM             1,636 CompoundInterest01.class
2022-06-15  04:01 PM             1,433 CompoundInterest01.java
               2 File(s)          3,069 bytes
               2 Dir(s)   672,338,796,544 bytes free

C:\PacktJavaCode\CompoundInterest01>java CompoundInterest01
If you deposit $100.00 in a savings account that pays 5% annual in
terest compounded monthly
you will have after 5 years $128.34
C:\PacktJavaCode\CompoundInterest01>|
```

Figure 2.8 – Running the Java program

Here, you can see the program's output.

On Windows systems, there is a second JVM loader called javaw that is used to execute a GUI application without opening a console. If you create a Windows shortcut to run a GUI Java program and use java.exe, then a console window will open, followed by the program's GUI window. If you use javaw.exe, the console window will not appear.

In most situations, using javac and then java is the most common way to work with Java code at the command line. A program that may consist of multiple files will require each file to be compiled, but only the file that contains the main method is executed. Let us look at one last way to compile and execute a Java program in a single step.

Launch Single-File Source-Code Programs

Before Java 11, the process of going from source code to execution was a two-step process: you compiled the code and then ran the code. Beginning with Java 11, another way to run a Java program was introduced, called **Launch Single-File Source-Code Programs**. This allows you to compile, start up the JVM, and execute the program in a single line. We'll see how this works for Windows, macOS, and Linux before examining a unique way for Linux and macOS.

For Windows, macOS, and Linux

Open a Command Prompt or Terminal in the same folder as the file you wish to run and, if necessary, update the Path and JAVA_HOME properties. Now, simply enter java and the name of the source file:

```
Command Prompt            ×    +  ∨                                   –    □    ×

Microsoft Windows [Version 10.0.22000.739]
(c) Microsoft Corporation. All rights reserved.

C:\PacktJavaCode\CompoundInterest01>java CompoundInterest01.java
If you deposit $100.00 in a savings account that pays 5% annual interest
compounded monthly
you will have after 5 years $128.34
C:\PacktJavaCode\CompoundInterest01>
```

Figure 2.9 – Running a Java program

As the name of this technique implies, your program can only consist of a single file. This source file may contain more than one class and the first class in the file must have a main method. Let us look at a new version of the program that is split into two classes in one file:

```
import java.text.NumberFormat;


public class CompoundInterest02 {
```

```java
    public static void main(String[] args) {
        var banker = new CompoundInterestCalculator02();
        banker.perform();
    }
}
```

The first class in this file just contains the main method, which is required. The second class, shown in the following code block, is in the same file; this is where the actual work is carried out:

```java
class CompoundInterestCalculator02 {
    private final double principal = 100.0;
    private final double annualInterestRate = 0.05;
    private final double compoundPerTimeUnit = 12.0;
    private final double time = 5.0; //
    private final NumberFormat currencyFormat;
    private final NumberFormat percentFormat;

    public CompoundInterestCalculator02() {
        currencyFormat =
                        NumberFormat.getCurrencyInstance();
        percentFormat = NumberFormat.getPercentInstance();
        percentFormat.setMinimumFractionDigits(0);
        percentFormat.setMaximumFractionDigits(5);
    }

    public void perform() {
        var result = calculateCompoundInterest();
        System.out.printf("If you deposit %s in a savings
                account " + "that pays %s annual interest
                compounded monthly%n" + "you will have
                after %1.0f years %s",
                currencyFormat.format(principal),
                percentFormat.format(annualInterestRate),
                time, currencyFormat.format(result));
    }
```

```
      private double calculateCompoundInterest() {
          var result = principal * Math.pow(1 +
                    annualInterestRate / compoundPerTimeUnit,
                    time * compoundPerTimeUnit);
          return result;
      }
}
```

The two classes are CompoundInterest02 and CompoundInterestCalculator02. In the Command Prompt or Terminal, enter java CompoundInterest02.java; you will get the same result:

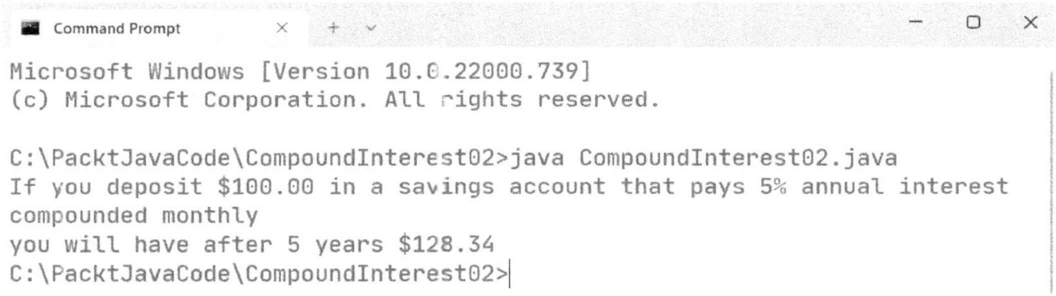

```
Command Prompt              ×    + ∨                                   —    □    ×
Microsoft Windows [Version 10.0.22000.739]
(c) Microsoft Corporation. All rights reserved.

C:\PacktJavaCode\CompoundInterest02>java CompoundInterest02.java
If you deposit $100.00 in a savings account that pays 5% annual interest
compounded monthly
you will have after 5 years $128.34
C:\PacktJavaCode\CompoundInterest02>
```

Figure 2.10 – Running a Java program with two classes in the java file

This technique does not create a bytecode .class file; it is only created in memory.

For macOS and Linux – Shebang files

There is a unique way to use Launch Single-File Source-Code Programs that is only available on macOS or Linux: **Shebang** files. These files can be run from the command line without needing to invoke the java command. This makes single-file Java programs usable as a script file akin to Bash. Shebang is the name given to the #! Characters.

Let's look at the beginning of the source code with the Shebang added:

1.   Add #! to the source code, as shown here:

```
#!//home/omniprof/jdk-17.0.3+7/bin/java --source 17
import java.text.NumberFormat;

public class CompoundInterest03 {
```

```
public static void main(String[] args) {
    var banker = new CompoundInterestCalculator03();
    banker.perform();
}
}
```

The first line that begins with the Shebang includes the path to the Java executable and the – source version switch. The version is the numbered version of Java you are using, which in this case is 17. To use this technique, the file must not have the .java extension. Rename the Java source code file to remove the extension.

2.   The next step is to make the file executable. Use chmod +x CompoundInterest03 to do so.

3.   Finally, you must execute the file by entering ./CompoundInterest03. Here is the output:

```
omniprof@Kona:~/CompoundInterest03$ ./CompoundInterest03
If you deposit $100.00 in a savings account that pays 5% annual interest compounded monthly
you will have after 5 years $128.34
omniprof@Kona:~/CompoundInterest03$ |
```

Figure 2.11 – Running a Java program with a Shebang in Linux

Here, we are running our Java program as if it were just a regular Linux or macOS program.

This concludes our topic on using Java from the command line. We started by looking at REPL in JShell, then the classic two-step approach, and ended with the Launch Single-File Source-Code Programs approach. We also covered the unique Shebang technique. Now, let us learn about the four most widely used IDEs.

Integrated development environments

It is time for a little honesty – very few Java developers work with just a text editor such as **vi** or **Notepad**. Knowing how to work with a standalone text editor and compile/execute at the command line is an important skill, but when given the choice of tooling, we will always go for an IDE. The features we will look at in this section will explain why this is so.

In this section, we will briefly review four of the most widely used IDEs available. Each IDE has a unique build system, which we will discuss in *Chapter 3, The Maven Build Tool*, and all the IDEs support the same external build systems. This means that in a team, each member can use the IDE that they feel makes them the most productive while being able to freely move code between team members without the need to make changes for a particular IDE. Before I introduce these IDEs, let us look at the features they all share.

Feature 1 – the code editor

The heart of every IDE is its editor. Like any ordinary text editor, it supports the usual list of features, such as cut, copy, and paste. What sets the IDE editor apart from these is that every keystroke is monitored. Should you mistype the name of a variable or method, you will immediately be informed on your screen of the error.

The editors also share JShell's ability to execute code one line at a time. This occurs out of sight. If you are executing code that generates an error – such as referring to a library that is not part of the project – you will be told of the error in the editor as you type rather than when you try to compile and execute the code. Most errors in your code, though not all, are detected as you type.

Another invaluable feature of these editors is called **code completion**. Microsoft calls this feature **IntelliSense**. Code completion can mean several things – for example, if you write an opening brace, bracket, or parenthesis, the IDE will add the closing one.

In Java, like some other OOP languages, the dot operator (.) indicates that you wish to call upon a member method or instance variable of an object. Code completion supports listing all the possible choices after the dot operator. The following figure shows all the choices for `percentFormat`:

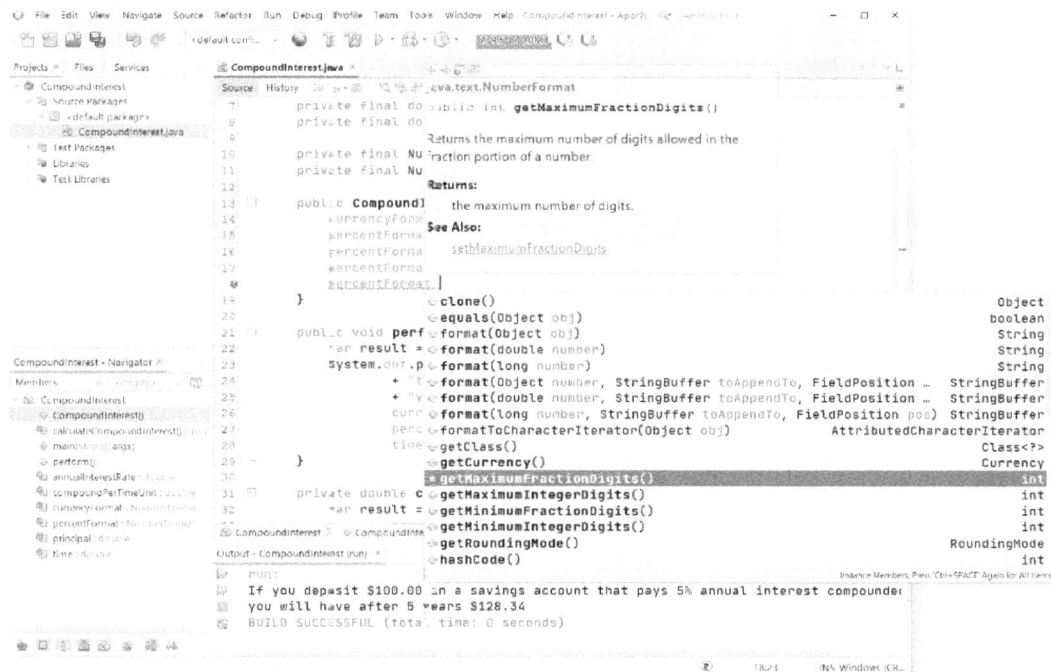

Figure 2.12 – Example of code completion in NetBeans

Code completion can also recommend changes in your code that are more efficient or modern. For example, the original Java **switch** statement is identical to the C switch. Recent enhancements to the switch syntax can effectively eliminate the ancient switch. If the IDE recognizes that the modern syntax could be used, then you will be advised and, with your permission, the IDE will rewrite your switch in the new format.

Feature 2 – server management

Servers represent external services that your code may need to interact with. Examples include database servers such as **MySQL** and **PostgreSQL** and web servers such as **Payara** and **WildFly**. From within the IDE, it is possible to stop and start these services. For database servers, you can connect the IDE to the server and write **SQL** queries and see the results. Applications or web servers can also be started and stopped. You can deploy or undeploy your compiled code to the server.

Feature 3 – debugging and profiling

Single-stepping through code is an invaluable feature of debugging tools. Modern IDEs provide such debugging capabilities; it is invaluable when your code runs but returns the wrong result. When running with a debugger, you can follow the execution of your program in the source code. You can inspect the state of the variables. Errors in the syntax of your coding are mostly recognized by the editor.

Profiling allows you to monitor your application while it runs. The profiler reports memory usage and the CPU time that a method consumes. This can be invaluable information in identifying where a program executes more slowly than you expected. Even if you did not suspect a problem with program speed, a profiler can provide you with the data you need to improve the performance of your program.

Feature 4 – source control management

Modern IDEs support interaction with source control management tools such as **Git**, **Mercurial**, and **Subversion**. These tools maintain your code in a repository. There is no need to use a standalone client tool to push or pull from a repository. Should a push result in a conflict, then the IDE can present what is currently in the repository and what you want to push and allow you to decide how to resolve the conflict.

Feature 5 – build systems

The only code we have seen so far has consisted of a single file. As we learn more about Java, we will discover that applications typically consist of multiple files. These files may be placed in multiple folders. Then, there are external libraries that provide capabilities that are not part of Java, such as the code necessary to interact with specific databases. A build system is responsible for ensuring that all the components and libraries are available. It is also responsible for running the Java compiler and then running the program.

All IDEs each have their own build system. The external build systems known as Apache Maven and Gradle, which are independent of an IDE, will be covered in the next chapter. The four IDEs we will cover in this chapter all support these external build systems. This means that if you write a program with IntelliJ that's been configured to use Maven, then the same files can be opened in NetBeans if it's been configured to use Maven.

Now, let us take a brief look at the four most widely used IDEs.

Eclipse Foundation – Eclipse

The Eclipse IDE was originally developed by IBM as a replacement for their existing Java IDE called VisualAge, which was written in Smalltalk. In 2001, IBM released the Eclipse platform, written in Java and released as an open source project. A board made up of companies working with Java was formed to oversee the development of Eclipse. As more and more companies joined the board, it was decided to create an independent open source organization. In 2004, the Eclipse Foundation was created, with the Eclipse IDE as its first open source project. Eclipse can run on Windows, macOS, and Linux.

You can download Eclipse from `https://www.eclipse.org/downloads/packages/`. There is the Eclipse IDE for Java developers for primarily desktop software development. A second version called the Eclipse IDE for enterprise Java and web developers adds support for server-side programming. Eclipse supports a range of plugins that add additional capabilities.

Let us see what the compound interest program – the one we wrote in *The first program* section – looks like in the Eclipse IDE:

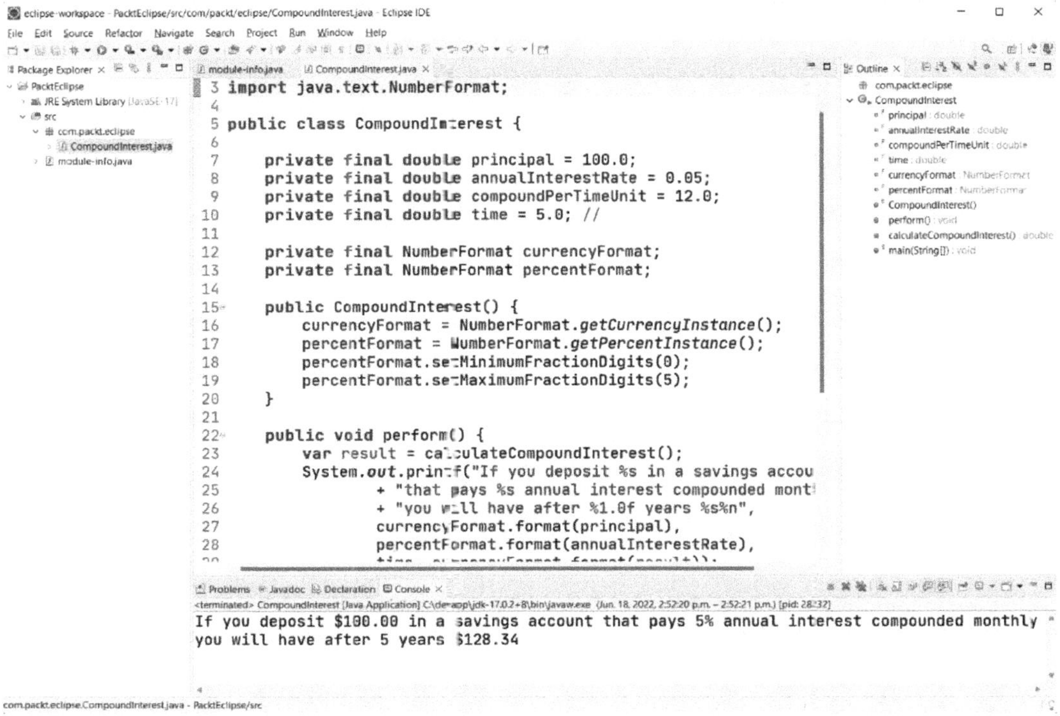

Figure 2.13 – The compound interest program in Eclipse

Here, you can see the output and what it looks like when you compile and execute it at the command line. All the IDEs capture the console output and display it in a window in the IDE.

Apache NetBeans

NetBeans began as a student project in 1996 in the Czech Republic. When James Gosling first encountered NetBeans while he was traveling to promote Java, he was so impressed by it that upon his return, he convinced the management at Sun Microsystems to buy the company behind NetBeans. In 2010, Oracle acquired Sun Microsystems, and in 2016, Oracle donated the NetBeans source code to the **Apache Foundation**. NetBeans runs on Windows, macOS, and Linux.

NetBeans has adopted an update cadence like Java, with new versions expected every 6 months. While not as feature-rich as some other IDEs, it is the simplest of the four to work with. As such, it is the ideal candidate when teaching Java or when you wish to use an IDE without a steep learning curve. As an Apache open source project, it is also the easiest to become involved with and contribute to.

You can download Apache NetBeans from `https://netbeans.apache.org/`. There is just one version and it supports desktop and server-side development. In addition, some plugins add additional functionality, such as support for frameworks such as Spring.

Let us see what the compound interest program – the one we wrote in *The first program* section – looks like in Apache NetBeans:

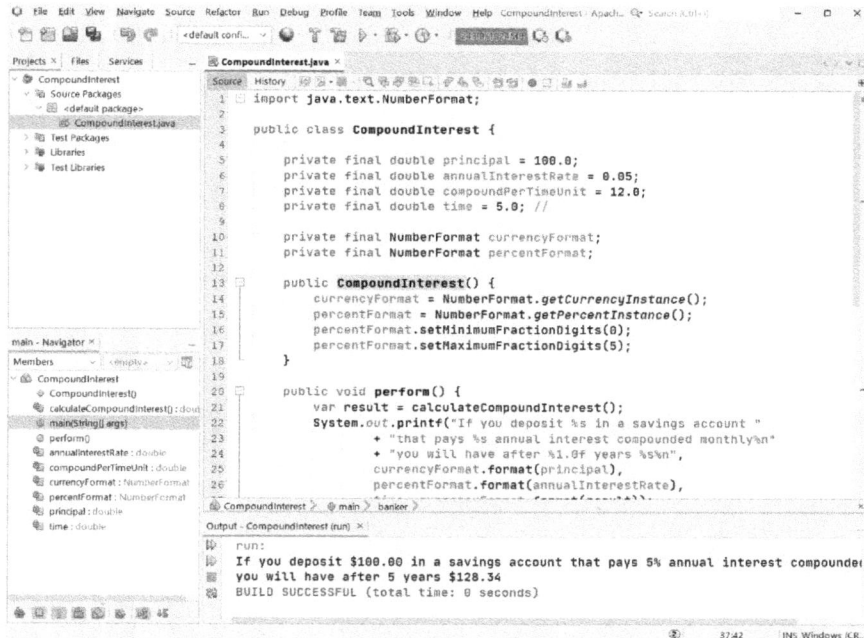

Figure 2.14 – The compound interest program in NetBeans

Here, you can see how NetBeans shows the output of the compound interest program.

Microsoft Visual Studio Code

Visual Studio Code (**VS Code**) was introduced by **Microsoft** in 2016. Its purpose was to be a development environment for a wide range of languages such as JavaScript, C++, Python, and Java, among others. The program consists of a core component that is released as open source. Support for specific languages is handled by extensions. The primary Java extension was developed by Red Hat, and unlike Microsoft-authored extensions, this one is open source.

VS Code is written in **TypeScript** and uses the open source Electron framework for creating desktop applications. VS Code is available for Windows, macOS, and Linux, though not all extensions work on every OS.

You can download VS Code along with several Java extensions from `https://code.visualstudio.com/docs/languages/java`. You will want to download the Coding Pack for Java, which contains VS Code plus the Java extensions. If you have already downloaded the basic version of VS Code, you can add Java support to an existing installation by downloading the Java Extension Pack.

Here's what the compound interest program – the one we wrote in *The first program* section – looks like in VS Code:

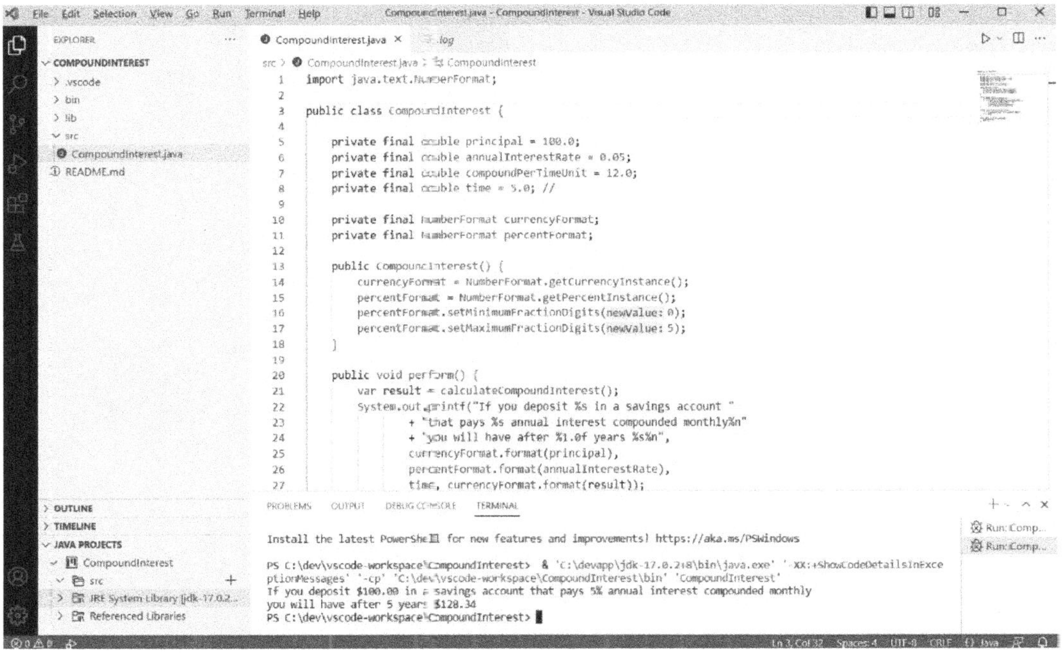

Figure 2.15 – The compound interest program in VS Code

Here, you can see how VS Code shows the output of the compound interest program.

JetBrains IntelliJ IDEA

IntelliJ IDEA from the JetBrains company, written in Java, was introduced in 2001. It comes in two flavors. First, there is a free Community edition with an open source license for developing desktop Java applications. A second commercial version, called Ultimate, includes support for additional Java frameworks such as Java EE/Jakarta EE and Spring. The commercial version requires an annual paid subscription.

IntelliJ is considered the most feature-rich of the Java IDEs. This does not necessarily mean it is the best, but it is the most widely used of all the IDEs. You can download it from `https://www.jetbrains.com/idea/download/`. As already mentioned, the Community edition is free, while the Ultimate version requires a subscription.

Let us see what the compound interest program – the one we wrote in *The first program* section – looks like in IntelliJ IDEA:

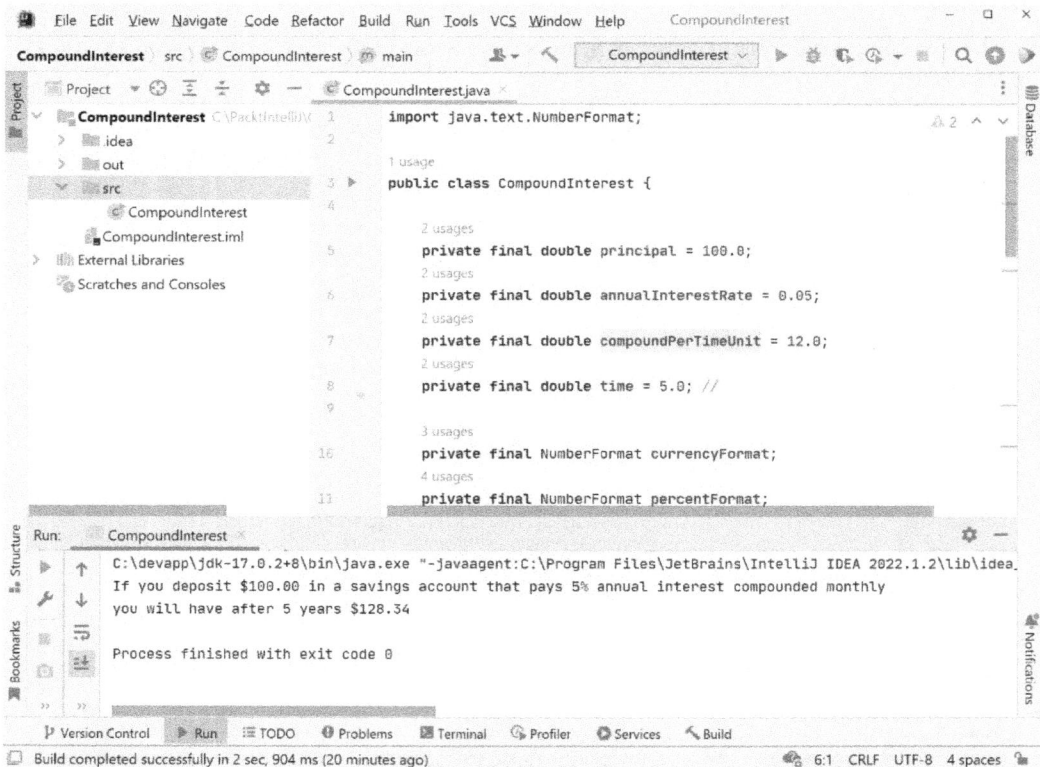

Figure 2.16 – The compound interest program in IntelliJ IDEA

Here, you can see the output in IntelliJ IDEA. This is what it looks like when you compile and execute at the command line.

Which IDE should you use?

There are two factors to consider when choosing an IDE:

- The first is whether the company you work for mandates a specific IDE. If so, then the choice has been made for you.
- The second is how you feel when using the IDE. All four of the IDEs shown here can support whatever you are coding in Java. They can also be used for other languages, such as C, C++, PHP, and JavaScript.

I can only explain what my choice was and the reasons for that choice. I needed an IDE that required the minimum amount of classroom instruction. I taught Java project courses in the final year of a 3-year computer science program at a college in Quebec. I needed to teach advanced desktop programming and introduce the students to server-side programming. I did not want to teach an IDE. For these reasons, I chose NetBeans. Students were permitted to use any of the other three IDEs if they used the external Maven build system, but if they ran into trouble with the IDE, I could only provide minimal assistance.

So, I recommend that you take the time to experiment with each of the IDEs. Select your personal IDE based on how you feel about using it. They are all effectively the same while presenting unique methods to get things done. All of this book's source code, which can be found in this book's GitHub repository, will run on all four IDEs mentioned here.

Summary

In this chapter, we examined the various ways in which we can write, compile, and execute Java programs from the command line. We looked at REPL in JShell to run snippets of code quickly. Then, we saw the classic way that Java is compiled and executed in two steps. Finally, we looked at Launch Single-File Source-Code Programs for executing Java programs written in a single file. With the Shebang concept found in macOS and Linux, we saw how Java could even be used as a scripting language. We ended by briefly looking at the four most common IDEs.

Now that you know how to write, compile, and execute Java programs, in the next chapter, we will explore an external build system that can be used from the command line or within an IDE. This topic will help explain why your choice of IDE is personal. You will see why developers can work together while members of the team may use a different IDE.

Further reading

To learn more about the topics that were covered in this chapter, take a look at the following resources:

- *Java Platform, Standard Edition – Java Shell User's Guide*: https://docs.oracle.com/javase/10/jshell/JSHEL.pdf

- *JEP 330: Launch Single-File Source-Code Programs*: https://openjdk.org/jeps/330

- *Eclipse Foundation*: https://www.eclipse.org/

- *Apache NetBeans*: https://netbeans.apache.org/

- *Visual Studio Code*: https://code.visualstudio.com/

- *JetBrains IntelliJ IDEA*: https://www.jetbrains.com/idea

3
The Maven Build Tool

Java programs are seldom just a single file. They can consist of just a few files or thousands of files. We have seen that you must compile Java source code files into bytecode. Having to do this for so many files makes the work quite tedious. This is where a build tool is invaluable.

In the previous chapter, the programs were all run from the folder we stored them in. As programs grow into multiple files, you manage them by categorizing them. The basic categories that date back to the early days of programming are input, process, and output. You can break down these categories into specific tasks that your program must perform. In Java, we call a category a **package**. A package, in turn, is a folder into which you store all the Java files that are part of the category. A complex program may consist of hundreds of files organized into packages.

In this environment, you must compile every single file. As you can imagine, this can be very tedious if you must compile them one at a time. One of the purposes of a build system is to simplify this task, and at the very minimum, all you must enter at the command line is mvn, the Maven executable program. In this chapter, we will see how we can use **Maven**. As Maven is a feature of every IDE, you can load any program organized as a Maven-managed project into any IDE.

> **Important note**
>
> While Maven is the most widely used build system, it is not the only one. Another popular build system is called **Gradle**. What sets it apart from Maven is that it uses an imperative rather than a declarative configuration file. Gradle uses a **Domain-Specific Language** (**DSL**) based on the **Groovy** language. As such, it could be used for general programming, although its vocabulary and syntax are designed for building software projects.

In this chapter, we will cover the following:

- Installing Maven
- Overview of Maven functionality
- The pom.xml configuration file
- Running Maven

By the end of this chapter, you will gain a sufficient understanding of the Maven build process to be able to use it right away. Later in the book, we will see how we can use Maven to manage testing. See the *Further reading* section for links to articles and free books that cover Maven in far more detail.

Technical requirements

Here are the tools required to run the examples in this chapter:

- Java 17 installed
- A text editor
- Maven 3.8.6 or a newer version installed

I suggest downloading the source code that goes with this book from `https://github.com/PacktPublishing/Transitioning-to-Java/tree/chapter03` before you proceed so that you try out what is shown in this chapter.

> **Note**
> Ubuntu and other Linux distributions may have a version of Maven already installed. If it is not version 3.8.6 or greater, you must replace it with the newest version.

Installing Maven

Visit the Maven download page at `https://maven.apache.org/download.html`. Here, you will find the program in two different compressed formats, one for Windows (`.zip`) and one for Linux/macOS (`tar.gz`).

Binary tar.gz archive	apache-maven-3.8.6-bin.tar.gz
Binary zip archive	apache-maven-3.8.6-bin.zip

Figure 3.1 – Maven compressed files

The versions shown here represent the current ones at the time of this writing. When starting out, it is best to install the most recent version. Now, let us review how we install Maven for each OS.

Windows

There is no installer for Maven. Unzip the ZIP archive into a folder. As we saw when we installed Java without an installer, I use a folder called `devapp` for all my development tools. Once unzipped, you need to add the location of the bin folder to your path. You may come across references to two environment variables, M2_HOME and MAVEN_HOME. While they don't do any harm, both are obsolete as of Maven 3.5.x.

If you are an admin on your computer, then just add the path to the bin folder to your path. If you are not, then use the set command to add it to your path. Here is my setjava.bat file for non-admins. Change your batch file to match your folder structure.

```
set JAVA_HOME=C:\devapp\jdk-17.0.2+8
set PATH=%JAVA_HOME%/bin;C:\devapp\apache-maven-3.8.6\
bin;%PATH%
```

You can verify that Maven is working with mvn --version:

```
C:\PacktJavaCode\CompoundInterest64>mvn --version
Apache Maven 3.8.6 (84538c9988a25eec085021c365c560670ad80f63)
Maven home: C:\devapp\apache-maven-3.8.6
Java version: 17.0.2, vendor: Eclipse Adoptium, runtime: C:\devapp\jdk-17.0.2+8
Default locale: en_CA, platform encoding: Cp1252
OS name: "windows 11", version: "10.0", arch: "amd64", family: "windows"
```

Figure 3.2 – The output of mvn --version on Windows

Linux

If you have superuser status, you can use this command:

```
$ sudo apt-get install maven
```

Verify it with mvn --version.

If you are not a superuser, unzip the tar.gz file downloaded from the Maven website into the folder of your choice:

```
$ tar -xvf apache-maven-3.8.6-bin.tar.gz -C /usr/local/apache-
maven/apache-maven-3.8.6
```

Now, add the location for Maven to your path:

```
export PATH=/usr/local/apache-maven/apache-maven-3.8.6/bin
:$PATH
```

Add this line to your .profile or .bash_profile file according to your Linux distribution.

macOS

Assuming that you are using Homebrew on your Mac, you can install Maven with the following:

```
$ brew install maven
```

Verify the installation with `mvn --version`.

If you do not have Homebrew or are not the superuser, then you can install Maven the same way as you installed it for Linux for non-superusers. Newer versions of macOS use `.zshenv` rather than `.profile` for your user script.

```
omniprof@Kong:~$ mvn --version
Apache Maven 3.8.6 (84538c9988a25aec085021c365c560670ad80f63)
Maven home: /home/omniprof/apache-maven-3.8.6
Java version: 17.0.3, vendor: Eclipse Adoptium, runtime: /home/omniprof/jdk-17.0.3+7
Default locale: en_US, platform encoding: UTF-8
OS name: "linux", version: "5.10.102.1-microsoft-standard-wsl2", arch: "amd64", family: "unix"
```

Figure 3.3 – Verifying the installation in Linux or macOS

With Maven installed, let us look at what it offers us.

Overview of Maven functionality

The standard Java libraries that are part of the JDK is quite extensive. However, there are added libraries that supply functionality, such as connecting to a relational database, which you must download and then add to the project before running it. You can configure Maven to do this for you. No need to visit a library's web page – download the file, place it in the proper folder, and let the Java compiler know it is available.

As with most build tools, Maven is more than just a tool for compiling programs.

In today's development environment, code does not go from the developer right into production if it compiles successfully. There has to be unit testing of methods and integration testing of the interactions between the various modules or classes in a program. You will use specialized servers for this work, and you can configure Maven to carry this out.

Let us review what else it can do.

Dependency management

While the Java language and its standard libraries can cover multiple use cases, they only cover a small subset of what programmers want to do. Frequently, added Java libraries that provide support for tasks, such as GUI programming with JavaFX; certain drivers, such as one for working with a range of databases, for sophisticated logging, or for enhanced data collections; and more, need to be part of your program. The problem is that these libraries must be in the **Java classpath**. A classpath is the list of files and folders holding Java libraries that must be accessible in your filesystem. Without a tool such as Maven, you must download every library you wish to use and update the Java classpath manually.

However, Maven allows you to list all the libraries you plan to use in a project's configuration file called the **Project Object Model (POM)**, with the file name pom.xml. Maven keeps a folder on your filesystem where it stores the required dependency library files. This folder is called a repository. By default, this folder is placed inside another folder called .m2, which, in turn, is stored in your home directory. You can change this to use any folder from your computer, although most programmers leave the default folder location as is.

If a required dependency is not already in your local repository, then Maven will download it. A default central repository called Maven Central exists and can be found at https://repo1.maven.org/maven2/. You can search for libraries and retrieve the entry necessary to add to your pom.xml file at https://search.maven.org/.

Maven plugins

The Maven program is not large; it relies on Java programs called plugins to conduct its tasks. For example, there are plugins to compile and package code, run tests, execute and deploy the code to a server, and more. The pom.xml file is where we list the plugins along with the dependencies. You can search for plugins as you did for dependencies at MVNRepository.

The Maven program uses a range of default plugins available to you without including them in the pom.xml file. Major builds of Maven use versions of the default plugins at the time the build was released. To ensure that you are using the most recent version of the plugins, I recommend listing every plugin that you will use in the pom.xml file. In general, always explicitly list the plugins you will use rather than allow Maven to use its built-in or implicit plugins.

Maven project layout

To use Maven, it is necessary to organize the program's folder into a specific layout. You can configure Maven to use a layout of your choice. What follows is the default layout. This will allow Maven to discover all the files and resources that are part of your project. Let us look at the folder structure for a desktop application:

```
Project Folder
    /src
        /main
            /java
            /resources
        /test
            /java
```

```
        /resources
  pom.xml file
```

You must create these folders and the pom.xml file before you start coding from the command line with an editor. Your IDE will create this structure if you indicate it is a Maven project when you create it. All IDEs will open a Maven project if it follows this folder layout.

After you have successfully built the program, you will find a new folder called target. This is where Maven stores the compiled source code files and the final packaged file called the jar file. Here is the folder structure of target:

```
/target
    /classes
    /generated-sources
    /generated-test-sources
    /maven-archiver
    /maven-status
    /test-classes
    project.jar file
```

Maven creates this folder when you build a project for the first time. Each build will replace any file in target with a newer version when you change the matching source code file. Never edit or change any files found in target because the next time you create a build, it will replace the files in target and any edits you made will be lost.

You can also instruct Maven to clean a project, and this results in contents of target being deleted. If the pom.xml file is instructing Maven to package your program as an archive, such as a JAR file, then you find the JAR or whichever archive you are creating in the target folder.

If you do not work with an IDE, you may consider writing a batch file or shell script to create this folder structure.

The next task is to create the pom.xml file with the required plugins listed – but before we do that, let us look at how we organize multiple source code files using packages.

Java source code packages

The Java languages encourages developers to organize their code based on functionality. This could be code to interact with a user, access records from a database, or perform business calculations. In this section, we will learn about packages and how to use them in a Maven project. We already know that you do not need a package. The first program, CompoundInterest, that we ran in the previous chapter, did not have any packages. This was handy when a project just consisted of a single file. Once a project encompasses multiple files, you will use packages.

As we are using Maven, the location of our packages must be `src/main/java`:

```
Project Folder
    /src
        /main
            /java
```

The rules for naming packages are similar to the rules for identifiers, as well as any rules for naming folders in your OS:

```
com.kenfogel.business
```

The periods stand for a slash, either a forward or back slash depending on your operating system. This means that `business` is a folder in `kenfogel` and `kenfogel` is a folder in `com`. Using our Maven layout, it will look as follows:

```
Project Folder
    /src
        /main
            /java
                /com
                    /kenfogel
                        /business
```

When we use packages, each file belonging to the package must have a statement as the first line of code that declares the name of the package.

In the previous chapter, we used a version of the `CompoundInterest` program with two classes in one file because the **Single-File-Source-Code** feature cannot have more than one file, as its name implies. Unless you need to use the Single-File-Source-Code feature, you should create a file for each class in your program. The file name must be the same as the public class name in the file.

Here is the class that holds the business process; notice that it begins with the package statement. In this first part, we are declaring what external library we will need and then declaring the class and the class variables:

```
package com.kenfogel.compoundinterest.business;

import java.text.NumberFormat;

public class CompoundInterestCalculator04 {
    private final double principal = 100.0;
```

```
    private final double annualInterestRate = 0.05;
    private final double compoundPerTimeUnit = 12.0;
    private final double time = 5.0;

    private final NumberFormat currencyFormat;
    private final NumberFormat percentFormat;
```

Next up are the methods that contain executable code:

```
    public CompoundInterestCalculator04() {
        currencyFormat =
            NumberFormat.getCurrencyInstance();
        percentFormat = NumberFormat.getPercentInstance();
        percentFormat.setMinimumFractionDigits(0);
        percentFormat.setMaximumFractionDigits(5);
    }

    public void perform() {
        var result = calculateCompoundInterest();
        System.out.printf(
        "If you deposit %s in a savings account " +
        "that pays %s annual interest compounded " +
        "monthly%nyou will have after %1.0f years " +
        "%s%n",
                currencyFormat.format(principal),
                percentFormat.format(annualInterestRate),
                time, currencyFormat.format(result));
    }

    private double calculateCompoundInterest() {
        var result = principal *
            Math.pow(1 + annualInterestRate /
                compoundPerTimeUnit,
                time * compoundPerTimeUnit);
        return result;
    }
}
```

Every Java program must have at least one class that has one method named main. Here is the class in which the main method exists:

```
package com.kenfogel.compoundinterest.app;

import com.kenfogel.compoundinterest04.business.
CompoundInterestCalculator04;

public class CompoundInterest04 {

    public static void main(String[] args) {
        var banker = new CompoundInterestCalculator04();
        banker.perform();
    }
}
```

In both files, you will see an import statement. To access a class that is not in the same package, you must import it. This statement informs the compiler that code coming from a class in another package will be used. Let's discuss both the import statements:

- The first import statement makes the NumberFormat class—that is part of the java.text package—available to the compiler. Note that packages that begin with java or javax are usually part of the Java installation:

  ```
  import java.text.NumberFormat;
  ```

- In the second import statement, by using CompoundInterest04.java, we are instantiating the CompoundInterestCalculator04 class in the main method. This class file is not in the same package, so you must import it to reference it:

  ```
  import com.kenfogel.compoundinterest.business.
                      CompoundInterestCalculator04;
  ```

Here is the folder structure of the CompoundInterest program that Maven expects to find on your storage device:

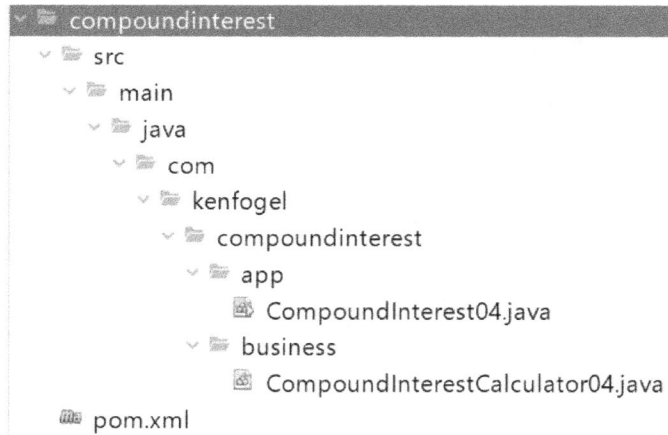

Figure 3.4 – Basic Maven file structure

Here, we are looking at the directory structure for the Maven-managed CompoundInterest program. The project is organized into packages that match import statements in your code. Here, you can see how the package names, such as com.kenfogel.compoundinterest, exist in your filesystem. We have one last part of a Maven project we must learn about and that is the pom.xml file.

The pom.xml configuration file

You must configure Maven with a declarative XML file. This file declares, in XML format, all the information necessary to build a project. It lists the required libraries here, along with the plugins needed to support Maven tasks. In this section, we will examine the pom.xml file that holds the Maven configuration to build the CompoundInterest program.

Here are the first three tags that are used in every pom.xml file by everyone:

```
<?xml version="1.0" encoding="UTF-8"?>
<project xmlns="http://maven.apache.org/POM/4.0.0"
    xmlns:xsi="http://www.w3.org/2001/XMLSchema-instance"
    xsi:schemaLocation="http://maven.apache.org/POM/4.0.0
    http://maven.apache.org/xsd/maven-4.0.0.xsd">

    <modelVersion>4.0.0</modelVersion>
```

Let's describe the parts of the code we just wrote:

- The first line that begins with <?xml is the XML prolog. These are the default values and could be left out, as it is optional.

- The root project tag defines the XML namespaces. It includes the location of the schema file that validates the pom.xml file.

- modelVersion refers to the version that the pom.xml file conforms to. Since Maven 2, it has been 4.0.0, and any other value will result in an error.

Next up is how we identify a project; consider the following code block:

```
<groupId>com.kenfogel</groupId>

<artifactId>compoundinterest</artifactId>

<version>0.1-SNAPSHOT</version>
```

These three tags are commonly referred to as the project's GAV, using the first letter of each tag – groupId, artifactId, and version. Taken together, these should be unique for every project you create.

groupId and artifactId also define a default package in your code. You do not need to have this package, but can have any package structure you think is appropriate. When a project, either your own or one downloaded for your project, is stored in your local repository, version becomes another folder. This allows you to have multiple versions of a project that differ by version number. Should you be developing a library for download by Maven, then the contents of the three tags become the identification for users to download your work.

You are free to use any names you want in groupId and artifactId. It must conform to the XML rules for a string. There is a convention that says that the combination of groupId and artifactId should conform to Java's rules for naming packages. The name should be unique, especially if you plan to make it available through Maven Central. Therefore, programmers use their company or personal domain name in reverse. If you do not have a domain, then simply use your name as I have. I do own kenfogel.com and I recommend that all developers get a domain name for their work.

For version, we are also free to use anything – numbers or strings. If you have previously downloaded a specific version of a library, then Maven will use the local copy. One special word, SNAPSHOT, when added to the end of the version designation, implies that this project is still in development. This means that Maven will download this library even if it exists in the local repository. Unless you reconfigure Maven, SNAPSHOT versions are only updated once a day.

Here is what Maven stores in your local repository after using Maven to run the project. You can see how it is stored using the complete GAV:

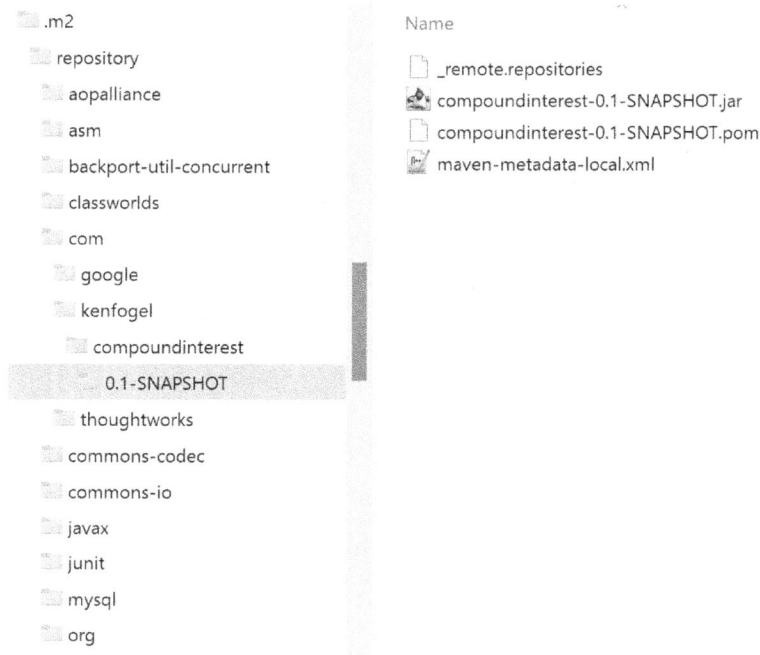

- .m2
  - repository
    - aopalliance
    - asm
    - backport-util-concurrent
    - classworlds
    - com
      - google
      - kenfogel
        - compoundinterest
          - 0.1-SNAPSHOT
      - thoughtworks
    - commons-codec
    - commons-io
    - javax
    - junit
    - mysql
    - org

Name
- _remote.repositories
- compoundinterest-0.1-SNAPSHOT.jar
- compoundinterest-0.1-SNAPSHOT.pom
- maven-metadata-local.xml

Figure 3.5 – The .m2/repository

The folders other than `compoundinterest` have dependencies and plugins that the Maven program and your project require.

Coming up, we will see a section of the pom.xml file called `defaultGoals` – a part of the `build` section of the pom.xml file. This is where you provide Maven with the tasks it must carry out. Maven does not place your project in your local repository unless you use `install` as one of your goals, and that is how this directory structure was created. The `groupId` element in the pom.xml file is broken down into folders based on the periods you placed in the tag. While `artifactId` and `version` in the pom.xml file may have periods in their text, they are not broken down into folders like `groupId`.

Next up is how we describe the final file that will contain everything necessary for your code to run. This is called `package` and refers to the various archive formats that Java programs can be stored in:

```
<packaging>jar</packaging>
```

If you do not have a <packaging> tag, then Maven will default to `jar`. These packages are compressed ZIP files with a folder structure and files required for the project. You can inspect any of these packaging formats by using any .zip utility. The choices for packaging are as follows:

jar – Java archive

Java archive is the basic ZIP archive file usually used for desktop applications. It is also usable for cloud-ready microservices. It contains a META-INF folder that has a file named MANIFEST.MF. If Maven has configured MANIFEST.MF to include the package and file name of the class that holds the main method, then you can run this file by double-clicking on it or entering the file name at the command prompt.

war – web archive

This ZIP archive is for use on a web server such as **Tomcat**. Tomcat is a widely used open source Java web server that supports the web profile. The file organization of the .war file differs from a .jar file to meet the requirements of a web server, such as folders for HTML and JavaScript files.

ear – enterprise archive

This ZIP archive is for use on a Java enterprise profile servers such as Glassfish or WildFly. These are also called application servers that provide the functionality for running complex web applications. Modern web programming in Java recommends using .war files even for complex systems. I will discuss these in more detail when we look at web programming in *Chapter 14, Server-Side Coding with Jakarta.*

pom – POM

Maven supports using multiple POM files. One way to do this is to have a parent POM file included as part of the project's POM file. I have hundreds of projects that I have created for my students. Early on, I found myself editing each POM file in each project to update versions or add new dependencies and plugins that every project will share. Using a parent POM file, I can place all the common components in this file and then include them in the individual POM files of each project. If the project and parent POM files each have the same tag, then the project POM overrides the parent POM.

Let us continue going through the pom.xml file:

```
<description>
    First example of a Maven multi source code project
</description>
<developers>
    <developer>
        <id></id>
        <name></name>
        <email></email>
    </developer>
```

```
        </developers>

        <organization>
            <name></name>
        </organization>
```

These are three optional sections that provide additional information that can help manage the project:

- `<description>`

  - Brief description of the project in sentences

- `<developers>`

  - Here you can list the team members. I have used it to identify my students.

- `<organization>`

  - Name of the company you work for or the name of the client

The `archive` file, the packaging that is created, holds all the compiled bytecode and any required libraries. It also includes the `pom.xml` file. This makes it possible for web and enterprise servers to display this information in the web console or dashboard.

Next up is the properties section of the file; consider the following code block:

```
<properties>
    <java.version>17</java.version>
    <project.build.sourceEncoding>
        UTF-8
    </project.build.sourceEncoding>
    <maven.compiler.release>
        ${java.version}
    </maven.compiler.source>
    <exec.mainClass>
        com.kenfogel.compoundinterest.app.CompoundInterest04
    </exec.mainClass>
</properties>
```

In the preceding code block, think of `properties` as variables that you can use elsewhere in the POM file, such as defining the Java version the compiler will come from or the name of the class containing the `main` method when creating a `MANIFEST.MF` file. You can see that `java.version` becomes

${java.version}. You can now use this value elsewhere in the POM file. The Maven plugin that manages compilation will use the compiler source and target source. exec.mainClass denotes the class that holds the main method.

Next up are dependencies; these are external libraries that your program requires. Consider the following code block:

```
<dependencies>

    <dependency>
        <groupId>org.openjfx</groupId>
        <artifactId>javafx-controls</artifactId>
        <version>18.0.1</version>
        <scope>compile</scope>
    </dependency>
    <dependency>
        <groupId>mysql</groupId>
        <artifactId>mysql-connector-java</artifactId>
        <version>8.0.29</version>
        <scope>runtime</scope>
    </dependency>
    <dependency>
        <groupId>org.junit.jupiter</groupId>
        <artifactId>junit-jupiter-engine</artifactId>
        <version>5.8.2</version>
        <scope>test</scope>
    </dependency>


</dependencies>
```

A dependency is a library that must be available for the program to compile and execute. If a dependency is not found in your local repository, then Maven will download it.

Just as you named your project the first three tags, groupId, artifactId, and version, name the library you wish to use. Maven uses this information to identify what it must look for in your local repository or a remote repository so it can be downloaded.

A new tag appears here called `<scope>`. Here are the four most used scopes:

- **Compile scope**: This is the default scope. It means that this library is required to compile the program. It will also be added to the package.

- **Runtime scope**: This library must be available at runtime, but it is not used for compiling. **Java Database Connectivity** drivers fall into this category, as Java only uses them when the program runs.

- **Provided scope**: When you run a program in a framework such as Spring or an application server such as WildFly, many of the project's dependency libraries are included in the server. This means you do not need to add them to the archive. You will need these files to compile the code, and Maven will download them into your repository so that the compiler can verify that you are using them correctly.

- **Test scope**: Unit testing involves the writing of test classes. These test classes are compiled and stored in the `test` branch of the Maven project and not the `java` branch.

The build section

Next up is the `build` section, where we define what tasks we want Maven to carry out and what we will need to accomplish. In Maven, you can express the tasks you want to carry out as either a lifecycle, a phase, or a goal. A lifecycle consists of multiple phases, a phase can consist of multiple goals, and a goal is a specific task.

In Maven, there are just three lifecycles, whereas there are numerous phases and goals:

```
<build>
    <defaultGoal>clean package exec:java</defaultGoal>
```

Here we have the `defaultGoal` tag of the `build` section of a POM file. If you do not use this tag, then Maven will use the `Default` lifecycle, which, in turn, invokes 21 phases. In this example, we are explicitly invoking two phases and one goal.

As the name of this tag implies, this is the set of phases and goals that will be performed in case no goals or phases were otherwise defined (via the command line). The `clean` belongs to the `Clean` lifecycle, which, in turn, consists of three phases. When we list a phase such as `clean`, Maven will also carry out every phase that precedes it. In the case of the `Clean` lifecycle, if you show the `clean` phase, then it also carries out the `pre-clean` phase but not the `post-clean` phase. To carry out all the operations of a lifecycle, you simply use the last phase of the lifecycle.

In this example, we see two phases and one goal. We just saw that the `clean` phase invokes its preceding phase first. The package phase is preceded by 16 phases, each of which will be carried out. A goal is a single task and does not invoke anything else. The `exec:java` goal is used to execute your code explicitly after all previous phases and goals complete successfully.

Here are a handful of the phases and goals we will be using.

Phases

- `clean`: Deletes the target folder. This will force Maven to compile all the source files. If not used, only source code files with a date and time later than the corresponding `.class` file are compiled. This goal is a member of the `Clean` lifecycle and does not invoke any other goals in other lifecycles.

- `compile`: `compile` will go through the source code tree, compile every source code file, and write the bytecode to the target folder. As a member of the `Default` lifecycle, and before Maven runs `compile`, it will first perform all the goals that precede it.

- `test`: This goal will invoke the unit tests. As the subsequent goals will run tests, we do not have to list them explicitly. However, if you just want to compile and test your code, then you can use `test` as the final goal.

- `package`: This combines all the files into a `jar` package assuming that the `<packaging>` tag in the POM file is `jar`. The test goal precedes the package in the `Default` lifecycle. Therefore, Maven will run the unit tests, if there are any, first.

- `install`: This adds this project to your local repository if all previous goals have been completed successfully.

Goals

- `exec:java` and `exec:exec`: These two are not part of the standard lifecycles. They require a special plugin and do not execute any other goals. `exec:java` will use the same JVM that Maven is running in. `exec:exec` will spawn or start up a new JVM. This can be useful if you need to configure the JVM.

You can override the `defaultGoal` tag by placing a phase name on the command line as follows:

```
mvn package
```

In this example, as the package belongs to the `Default` lifecycle, all phases that precede it will be carried out first. All the phases and goals in `defaultGoal` are ignored.

Plugins

Here is the concluding section of the build, where we will define the plugins. Except for `maven-clean-plugin` and `exec-maven-plugin`, all of these plugins exist in Maven as defaults. The versions of the plugins are decided when a major revision occurs, such as going from Maven 2 to Maven 3. This list is not updated with point releases.

Maven 3, introduced in 2010, has an internal list of default plugins that is quite old. For that reason, you should declare every plugin you will use, even if there is a default; that is what you see here.

Some plugins have tags that allow you to configure how they carry out their task. `maven-jar-plugin` allows you to show the class that holds the main method in its `<mainClass>` tag. We will configure the `surefire` plugin when we examine unit testing to turn unit tests on or off. As we look at different programs, we will be enhancing this and other POM files we will use:

- This plugin is responsible for deleting any output from a previous run of Maven:

```
<plugin>
    <groupId>org.apache.maven.plugins</groupId>
    <artifactId>maven-clean-plugin</artifactId>
    <version>3.2.0</version>
</plugin>
```

- This plugin includes any files in the resource folder of a project into the final packaging; a resource can be an image or a properties file:

```
<plugin>
    <groupId>org.apache.maven.plugins</groupId>
    <artifactId>
        maven-resources-plugin
    </artifactId>
    <version>3.2.0</version>
</plugin>
```

- This is the plugin that invokes the Java compiler:

```
<plugin>
    <groupId>org.apache.maven.plugins</groupId>
    <artifactId>
        maven-compiler-plugin
    </artifactId>
    <version>3.10.1</version>
</plugin>
```

- If you are performing unit tests, this plugin is used to configure the tests, such as to write the test results to a file:

```
<plugin>
    <groupId>org.apache.maven.plugins</groupId>
    <artifactId>
        maven-surefire-plugin
```

```
        </artifactId>
        <version>2.22.2</version>
    </plugin>
```

- This is the plugin responsible for packaging your program into a jar file. It includes the configuration to make the jar file executable by just double-clicking on it:

```
<plugin>
    <groupId>org.apache.maven.plugins</groupId>
    <artifactId>maven-jar-plugin</artifactId>
    <version>3.2.2</version>
    <configuration>
        <archive>
            <manifest>
                <mainClass>
                    ${exec.mainClass}
                </mainClass>
            </manifest>
        </archive>
    </configuration>
</plugin>
```

- This plugin allows Maven to execute your program:

```
<plugin>
    <groupId>org.codehaus.mojo</groupId>
    <artifactId>exec-maven-plugin</artifactId>
    <version>3.0.0</version>
</plugin>
        </plugins>
    </build>
```

- We close the root tag:

```
</project>
```

With our POM file ready, we are now ready to use Maven.

Running Maven

Once you have set up the Maven file structure, written your pom.xml file, coded your sources, and added any resources, such as images, then all you need to do is use Maven, which is quite straightforward.

Let us begin by running Maven on the command line.

Command-line Maven

Here are the steps to follow to use Maven on the command line:

1. Open a terminal or console in the folder that holds the project's folders, such as src.

2. Configure your setup if needed, should you not be an admin or superuser.

3. Enter the mvn command at the prompt. If there are no errors in your code, it should conduct all the goals you asked for. If there are errors, then you need to review the output of Maven, correct the errors, and use mvn again.

Here is my output from a successful build:

```
C:\PacktJavaCode\CompoundInterest04>mvn
[INFO] Scanning for projects...
[INFO]
[INFO] -----------------< com.kenfogel:compoundinterest >----
----------------
[INFO] Building compoundinterest 0.1-SNAPSHOT
[INFO] --------------------------------[ jar ]----------------
----------------
```

First up is to clean, meaning delete, any code generated the last time we built this program:

```
[INFO]
[INFO] --- maven-clean-plugin:3.2.0:clean (default-clean) @
compoundinterest ---
[INFO] Deleting C:\PacktJavaCode\CompoundInterest04\target
```

If there were resources, then we would be able see that they were added to the program. We do not have resources, so this plugin will do nothing:

```
[INFO]
[INFO] --- maven-resources-plugin:3.2.0:resources (default-
resources) @ compoundinterest ---
[INFO] Using 'UTF-8' encoding to copy filtered resources.
```

```
[INFO] Using 'UTF-8' encoding to copy filtered properties
files.
[INFO] skip non existing resourceDirectory C:\PacktJavaCode\
CompoundInterest04\src\main\resources
```

Now, the compiler is invoked. As we first cleaned the project, the plugin detected that all source code files must be compiled. If we did not use the clear goal, it would only compile source code files whose date is more recent than the byte-code file:

```
[INFO]
[INFO] --- maven-compiler-plugin:3.10.1:compile (default-
compile) @ compoundinterest ---
[INFO] Changes detected - recompiling the module!
[INFO] Compiling 2 source files to C:\PacktJavaCode\
CompoundInterest04\target\classes
```

You can have resources that are only used for unit testing. If there are any, they will be added to the test build of the project:

```
[INFO]
[INFO] --- maven-resources-plugin:3.2.0:testResources (default-
testResources) @ compoundinterest ---
[INFO] Using 'UTF-8' encoding to copy filtered resources.
[INFO] Using 'UTF-8' encoding to copy filtered properties
files.
[INFO] skip non existing resourceDirectory C:\PacktJavaCode\
CompoundInterest04\src\test\resources
```

The compiler is now invoked a second time to compile any unit test classes you have written:

```
[INFO]
[INFO] --- maven-compiler-plugin:3.10.1:testCompile (default-
testCompile) @ compoundinterest ---
[INFO] No sources to compile
```

This plugin is responsible for running the unit tests that were just compiled:

```
[INFO]
[INFO] --- maven-surefire-plugin:2.22.2:test (default-test) @
compoundinterest ---
[INFO] No tests to run.
```

As the packaging was defined as `.jar`, this plugin will now create the `.jar` file:

```
[INFO]
[INFO] --- maven-jar-plugin:3.2.2:jar (default-jar) @
compoundinterest ---
[INFO] Building jar: C:\PacktJavaCode\CompoundInterest04\
target\compoundinterest-0.1-SNAPSHOT.jar
```

This last plugin will execute your code:

```
[INFO]
[INFO] --- exec-maven-plugin:3.0.0:java (default-cli) @
compoundinterest ---
```

Here is the program output:

```
If you deposit $100.00 in a savings account that pays 5% annual
interest compounded monthly,
you will have after 5 years $128.34
```

All has gone well, and you receive the following report on how long the whole process took:

```
[INFO] -----------------------------------------------------------
----------------
[INFO] BUILD SUCCESS
[INFO] -----------------------------------------------------------
----------------
[INFO] Total time:  2.982 s
[INFO] Finished at: 2022-07-10T13:27:19-04:00
[INFO] -----------------------------------------------------------
----------------
```

Java requires that every statement or expression must end with a semicolon. I purposely removed a semicolon from one of the files so that we can see how coding errors are expressed. The following appears after a build failure is declared:

```
[ERROR] Failed to execute goal org.apache.maven.plugins:maven-
compiler-plugin:3.10.1:compile (default-compile) on project
compoundinterest: Compilation failure
[ERROR] /C:/PacktJavaCode/CompoundInterest04/src/
main/java/com/kenfogel/compoundinterest/business/
CompoundInterestCalculator04.java:[31,53] ';' expected
```

You can also request more information on failures by running Maven as mvn -X. This will provide more information should the error be due to a problem with the pom.xml file.

Running Maven in an IDE

Maven is normally included in IDE distributions. If you do not plan to work from the command line, you do not need to download and install Maven unless your IDE requests.

All IDEs have a run command and/or a run maven command. If both exist, use run maven. If there is no run maven command, expect the run command to recognize that this is a Maven project and use Maven rather than its internal build system to process your program.

An IDE will highlight errors in your source code and pom.xml file before you run the project. When the IDE recognizes errors, it will not compile your code until the issues are resolved.

Summary

In this chapter, we have learned how to use Maven, the most widely used Java build tool. The heart of Maven is the pom.xml file; we have seen the most significant sections of this file and what they are used for. Going forward, all the examples will be Maven-based.

By now, you know how to organize the directories for a Maven project, the components of a basic pom.xml file, and how to invoke Maven to build and execute your program.

Next, we will examine the object-oriented structure of a Java program, what an object is, and the coding syntax for loops and decisions.

Further reading

- *Maven: The Complete Reference*: https://books.sonatype.com/mvnref-book/reference/index.html

- *Maven by Example*: https://books.sonatype.com/mvnex-book/reference/index.html

- *The Maven Cookbook*: https://books.sonatype.com/mcookbook/reference/index.html

Part 2:
Language Fundamentals

You are an experienced coder who needs to learn the syntax of the Java language as quickly as possible. This part of the book covers the details you need to know to construct and code a solution in Java.

This part contains the following chapters:

- Chapter 4, *Language Fundamentals – Data Types and Variables*
- Chapter 5, *Language Fundamentals – Classes*
- Chapter 6, *Methods, Interfaces, Records, and Their Relationships*
- Chapter 7, *Java Syntax and Exceptions*
- Chapter 8, *Arrays, Collections, Generics, Functions, and Streams*
- Chapter 9, *Using Threads in Java*
- Chapter 10, *Implementing Software Design Principles and Patterns in Java*
- Chapter 11, *Documentation and Logging*
- Chapter 12, *BigDecimal and Unit Testing*

4

Language Fundamentals – Data Types and Variables

Now that we are comfortable (I hope) with basic Java tooling, we are ready to look at the language itself. As you are already a developer, there is no need to cover low-level concepts, such as what a variable is, in this chapter. So, this chapter will take advantage of what you already know and introduce you to the data types available in Java and the operations we can perform on them.

In this chapter, you will learn about the following:

- Type safety
- The eight primitive data types
- Literal values
- The `String` data type
- Naming identifiers
- Constants
- Operations on data
- Casting
- Overflow and underflow
- The math class

Technical requirements

Here are the tools required to run the examples in this chapter:

- Java 17
- Text Editor
- Maven 3.8.6 or a newer version

You can find the code from this chapter in the GitHub repo at `https://github.com/PacktPublishing/Transitioning-to-Java/tree/chapter04`.

> **Note**
>
> Ubuntu and other Linux distributions may have a version of Maven already installed. If it is not version 3.8.6 or greater, you must replace it with the newest version.

Primitive data types

Primitive data types create value variables. This means that once you declare variables in a program, you can use them in your code. However, before being represented by reference variables, classes must be instantiated into objects. But the values do not need to be instantiated.

In the `CompoundInterest` program, we needed to instantiate the `CompoundInterestCalculation` class before we can use it, as shown here:

```
var banker = new CompoundInterestCalculator04();
```

On the other hand, when we needed variables to hold `principal`, `annualInterestRate`, `compoundPerTimeUnit`, and `time`, we simply declared them, as shown in the following code line – we are directly assigning a value to the variable. We did not add the new operator, which is responsible for converting classes into objects. Primitive data types are ready to go:

```
double principal = 100.0;
```

There are eight primitive types in Java. Before we look at them, let us quickly see what type safety means.

Type safety

Depending on the language you are coming from on your path to Java, the concept of **type safety** may or may not be something you are familiar with. One form of type safety means that every variable must show its type when declared, and this type cannot change. You cannot assign an integer variable to a string. If you do this, you will get an error message as an exception. This is static typing.

The alternative to static typing is dynamic typing. Here, it is not necessary to declare the type of the variable. Java infers the type from what you assign. It is frequently and incorrectly assumed that dynamic typing is not type-safe. This is not necessarily true.

How you declare a variable is not at the heart of type safety. Instead, it is how the language handles, at runtime, what happens if a variable's type does not match the type required.

Here is a Python script that demonstrates that Python is type-safe even though it uses dynamic typing:

```
def print_hi(name):
```

```
    name = name + 2
    print(f'Hi, {name}')



if __name__ == '__main__':
    x="bob"
    print_hi(x)
```

In this example, the print_hi function is expecting to receive a variable named name. The first line of code in this function performs a math operation using the name variable.

In the code that is calling print_hi, we are declaring a variable, x, as a string. We know this because we are assigning a string to it. In this small snippet of code, it should appear obvious that this will generate an error. It does, and here is the error:

```
C:\devapp\PycharmProjects\PythonTest\venv\Scripts\python.exe
C:/devapp/PycharmProjects/PythonTest/main.py
Traceback (most recent call last):
  File "C:\devapp\PycharmProjects\PythonTest\main.py", line 8,
in <module>
    print_hi(x)
  File "C:\devapp\PycharmProjects\PythonTest\main.py", line 2,
in print_hi
    name = name + 2
TypeError: can only concatenate str (not "int") to str


Process finished with exit code 1
```

Python only detects this problem at runtime, but it is an error that will end the program. This means that despite the argument that dynamically typed languages are not type-safe, we have just seen that this is not the case. Python is effectively type-safe.

Java, on the other hand, is a statically typed language. Here is the same code in Java:

```
public class TypeSafetyTest {

    private void print_hi(String x) {
        System.out.printf(x);
    }


    public void perform() {
```

```
        int x = 4;
        print_hi(x);
    }


    public static void main(String[] args) {
        var typeTest = new TypeSafetyTest();
        typeTest.perform();
    }
}
```

Notice that print_hi clearly expects a string, but in the perform method, we are passing an integer. When we run this code with Maven, we will get the following error message:

com/kenfogel/typesafetytest/TypeSafetyTest.java:[12,18]
incompatible types: int cannot be converted to java.lang.String

In Python, you determine the variable type by where and how you use it. The Python print_hi method does not indicate the type of name. Only when we see the math expression in the function do we recognize that name must be an integer to work. The statically typed Java, by virtue of requiring the type as part of every declaration, makes it easier to spot type errors.

While we are comparing to Python – an excellent language – note that there is a significant difference in each language's compiler. The Java compiler can incrementally compile a program. The editor in an IDE can detect errors while you type by compiling the code one line at a time. Languages with no incremental compiler, such as Python, can only report errors in an IDE when you compile or run the code.

Which approach is better? I leave that up to you to decide.

However, static typing leads to more verbose programs. This means you must enter more code in a Java program compared to Python. On the other hand, static typing makes tracking down type errors easier and enhances the readability of the code.

One last point about Python – in version 3, the language developers introduced type annotations. These appear in Python code as if you are statically typing a variable. But this is not the case, as the compiler ignores these annotations. They exist to support type checkers, such as what the **PyCharm** IDE can do.

Before we look at the data types we use when declaring variables, let us take a moment to look at literal values.

Literal values

A **literal value** is one you enter into source code as a value, not a variable. Java treats a literal number as an integer if the number does not have any decimal places. This means that when you write a literal such as 42, Java treats this as an integer. Should the literal value exceed the range of an integer when assigned to an integer, you will get an `integer number to large` compiler error. If you assign the literal value to a `long` integer, you must add the letter L to the number, such as 14960000000000L.

When we write large numbers, we frequently use a separator every three digits to improve legibility. If you want a separator to make the source code easier to read, you can only use the underscore. You cannot use a comma or any other character as a separator. The value 14960000000000L can be entered as 14_960_000_000_000L.

When working with literal floating-point numbers, the default primitive type is double. If you are assigning a literal double that does not have any decimal places, then add one, rather than adding it like so:

```
double value = 100;
```

Enter it in the following manner:

```
double value = 100.0;
```

Alternatively, you could use the suffix D for double and F for float. You can write all literal suffixes in upper or lowercase, as shown here:

```
double value = 100D;
```

Now, let us move on to the primitive data types.

Integers

There are four members of the integer family – `byte`, `short`, `int`, and `long`. The difference is the number of bytes they use to contain a value. Java, like most languages, encodes integers using two's complement. This means that the range of values for any of the integer types goes from negative to positive values. Like Python, Java does not have unsigned integers, while C, C++, and C# do.

In the following table, you will find the size in bytes, the allowable range, and how to declare, assign, or declare and initialize all the members of the integer family:

Variable	Size	Range	Declaration	Assignment	Declaration & initialization
byte	1	-127 to +128	byte value;	value = 65;	byte value = 65;
short	2	-32_768 to +32_767	short value;	value = 498;	short value = 498;
int	4	2_147_483_648 to +2_147_483_647	int value;	value = 56_334;	value = 56_334;
long	8	-9_223_372_036_854_775_808 to +9_223_372_036_854_775_807	long value;	value = 14_960_000_000_000L;	long value = 14_960_000_000_000L;

Table 4.1 – Specs for integers and how we can use them

The integer data type on most computers is related to the size of a CPU's registers. The JVM is a 32-bit virtual machine, meaning that its registers are 32 bits or 4 bytes wide. While the JVM program is implemented as a 64-bit application, it remains an implementation of a 32-bit computer.

Floating point

Like most languages, Java uses a subset of the **IEEE standard binary floating-point numbers** to represent a floating-point value in memory. In the *Further reading* section, you can find links to websites that delve into this physical format. From our perspective, the main interest we have is in accuracy.

Accuracy is defined as representing a value exactly. We describe integers as accurate because every decimal integer number can be converted into a binary number. We call the conversion a **lossless conversion**.

Not all decimal floating-point values map to a fixed-length binary value, though. One of the best examples is 0.1 in decimal. This is 1 divided by 10. If we divide binary 1 (1) by binary 10 (1010), the result will be an infinitely repeating sequence of 0.00110011001100110011 . . . What this means is that floating point does not have the same accuracy as integers. The IEEE 754 standard deals with this issue, but you must always know that floating-point values are approximations. We call this a **lossy conversion**. How accurate or approximate they are is related to the two types of Java's floating-point data types – float and double. The measure of accuracy when we convert from decimal to binary floating point and back is referred to as the precision of the result. If a number exceeds the precision, it is considered an approximation of the actual result.

In the following table, you will find the size in bytes, the allowable range, and how to declare, assign, or declare and initialize all the members of the floating-point family. This information is critical in deciding whether to use double or float:

Variable	Size	Range	Precision	Declaration	Assignment	Declaration & initialization
float	4	±3.40282347E+38F	7	float value;	value = 2.4F;	float value = 2.4F;
double	8	±1.79769313486231570E+308	15	double value;	value = 120.234;	double value = 120.234;

Table 4.2 – Specs for floating point and how we can use it

We usually interpret precision as the number of valid numbers to the right of the decimal place. Float uses 23 bits, and double uses 53 bits. Therefore, thinking in terms of the number of digits is a crude way to define precision; it is about the length of the mantissa as per the IEEE 754 standard. Simply put, doubles, by virtue of their larger mantissa, have a larger range of values and a higher level of precision than float.

You may now think you should only be using double rather than using float. After all, we all want our results to be as accurate as possible. But the fact that doubles are twice the size in bytes, 64 as opposed to 32, has a performance penalty. In deciding whether to use double or float, consider the range of values and the precision required. For example, the float will be sufficient if the range of values is small and the number of decimal places after the decimal point will never exceed approximately six. The math operations you may perform will also influence your choice. Addition and subtraction are not concerns, but multiplication and division may have an impact.

The Java compiler can recognize when you assign a floating-point value to an integer variable. You will get a possible lossy conversion from double to int error if you do. Later in this chapter, we will look at casting to convert from one numeric data type to another.

We are now finished with the numeric types – integers and floating points. Now, let's look at the non-numeric types.

Boolean

The **Boolean** data type represents the value of a single bit – either zero or one. Zero means false and one means true. In Java, the set of values that you can assign to a Boolean is the true and false keywords. The result of all logical operations, such as *Is x greater than y?*, expressed as x > y, always returns a Boolean value.

In Python, you can cast an integer to or from a Boolean value. The C language does not have a Boolean type, so the language uses the integer values of zero for false and not zero for true. C++ has a Boolean type, but it is just a subset of integers with the zero and one values represented by the true and false keywords. C++ treats integers as Booleans, the same way C does.

In Java, a Boolean is a distinct type. You cannot use an integer in place of either true or false. This means you cannot use the result of a calculation that may be either zero or not zero, where you require a Boolean type.

In the following table, you will find the set of allowable values and how to declare, assign, declare, and initialize Booleans. Technically, you only need a single bit to represent true or false. However, there is no machine language or bytecode instruction that can read just one bit. The Java language architects have left the size in bytes of a Boolean up to the specific implementation of Java:

Variable	Boolean
Size in bytes	Implementation dependent
Range	true, false
Declaration	Boolean value;
Assignment	value = false;
Declaration and initialization	Boolean value = true;

Table 4.3 – Specs for Boolean and how we can use it

The CPU retrieves data from memory in units of bytes, typically 4 bytes at a time, as this is the word size of the CPU. It cannot directly read a single bit in RAM. Once retrieved and stored in a CPU register, the CPU can determine the state of any bit in a byte. This means that a Boolean can be no smaller than a byte. Java does not define the number of bytes the same way integers and floating points are. The number of bytes a Boolean uses depends on the implementation of the virtual machine. This can mean that the implementation of Java by one organization may use a different number of bytes compared to another.

Booleans are at the heart of decision-making and iteration for many organizations. Now, let's move on to the data type used to represent the characters of our written language.

char

The char data type contains the numeric value for 2-byte Unicode characters. Unicode UTF-8 is a variable-length character encoding from 2 to 4 bytes per character. Currently, Java only supports 2-byte encoding. The first 128 characters are identical to the first 128 characters found in ASCII encoding.

In C and C++, a char is a subset of integers, and you can use it as an integer. Python does not have a character type but uses strings with a length of 1 for a single character. In Java, a char is a unique data type; you cannot use it for an integer as C and C++ allow. You can cast a char to an integer or cast an integer to a char.

Take note that a single quotation mark around a single character implies a char.

In the following table, you will find the set of allowable values and how to declare, assign, or declare and initialize a char:

Variable	char
Size in bytes	2 bytes
Allowable values	0 to 65,535 UTF-8 character codes
Declaration	char letter;
Assignment	letter = 'Z';
Declaration and initialization	char letter = 'A';

Table 4.4 – Specs for char and how we can use it

There is one more table to look at and that is the default value assigned to variables that are not initialized when declared. Variables can be declared as fields in a class or as local variables in a method. Here are the default values for fields:

Type	Default value
boolean	false
byte	0
short	0
int	0
long	0
float	0.0f
double	0.0d
char	\u0000 (Unicode equivalent to null)
Reference to objects	null

Table 4.5 – Default values for fields

There is no default value for variables declared in a method. Any code that tries to read a local variable that has not been assigned an initial value will result in a compile-time error.

We have now gone over the eight primitive types. There is one more type that you can use, similar to a primitive, but it is not primitive. Let's meet String.

A special case – String

A string, with a *lowercase s*, is a list of characters that usually represent a word we may write or speak. String, with a *capital S*, is a class that contains a list of zero or more characters and numerous operations that you can perform on them. As a class, it normally must be instantiated into an object. As developers commonly use String objects, Java can perform the instantiation implicitly whenever you use the assignment operator (=) with a String variable. When referring to this data type, we always capitalize the first letter. This way, we know that we are referring to the String class. We will cover classes and objects in more depth in the coming chapter.

Let us examine `String` and how we use it. We begin with the specification table:

Variable	String
Size in bytes	Reference: 4 bytes Length: available memory
Allowable values	0 or more UTF-8 character codes
Declaration	String name;
Assignment/Instantiation	name = "Dog"; name = new String("Dog");
Declaration and initialization	String name = "Dog"; String name = new String("Dog");

Table 4.6 – Specs for String and how we can use it

In this table, **Size in bytes** refers to two parts of `String`. The first, called a **reference**, is a variable that contains the address in memory of the `String` object. A reference is like a pointer in other languages, but you cannot manipulate it as you can in C or C++. The **length** of this object in memory includes overhead for the object in addition to the characters in the actual text you are storing.

Classes become objects in memory by using the `new` operator in Java, as shown in the previous table. Developers use `String` objects frequently, and Java simplifies its usage by implicitly instantiating it when assigned a value. You can use `new` as shown, but this is rarely written this way. Instead, `String` appears to work like a primitive value for the convenience of programmers.

With the eight primitive types defined along with the one special case, we can now move on to how we can use them in our code.

Naming identifiers

An **identifier** in any language is simply the name we assign to a variable, class, or method. We will first focus on naming variables, and in *Chapter 5, Language Fundamentals – Classes*, we will look at naming classes and methods.

There are very few rules in Java related to naming identifiers, but for those that are, the compiler enforces them. These are as follows:

- The first character of an identifier can be one of the following:

  - Dollar sign ($)

  - Underscore (_)

  - Alpha character (A-Z, a-z)

- Subsequent characters can be any of the previously mentioned ones and numbers.

Once you adhere to the rules, the choice of naming is up to you. This is because Java has conventions for naming. A convention is not a rule, and the compiler does not validate them. Instead, conventions are techniques the programming community recommends for a given language. While working in a team, your fellow members expect you to follow these conventions. Here are the conventions for naming variables:

- The name of a variable should be a noun; variables are things and not actions.

- The first character should be lowercase. The convention for class identifiers requires that its first character be a capital letter.

- When using a name made up of more than one word, use camel case. Each word in the identifier should be lowercase except for the first character. Each subsequent word in the identifier after the first word must begin with a capital letter. It is acceptable if you prefer using the underscore to represent a space in a multi-word identifier rather than camel case. For this usage, all characters should be lowercase.

- Do not use abbreviations; use whole words.

- Avoid single-character identifiers. There are a limited number of cases where a single character is acceptable, such as for a loop index variable. Otherwise, use meaningful names.

- This table describes the naming conventions for a variable:

Convention	Acceptable	Unacceptable
Noun	double salary;	double receive;
First character lowercase	int cars;	int Cars;
Camel case	int platesOfPasta;	int platesofpasta;
Underscore separator	int plates_of_pasta;	int plates_Of_Pasta;
Abbreviations	Never acceptable	int lol; representing layers of lacquer

Table 4.7 – Naming conventions

- You should not use the dollar sign; the compiler uses it to create identifiers.

- You should not use the underscore as the first character as in other languages, such as C++, because it means the same as the dollar sign as the first character in Java.

- Once past the first character, you can use any letter of the alphabet, any number, an underscore, or the dollar sign.

Now let us look at the length of identifiers compared to other languages in the following table:

Language	Maximum # of significant characters
Python	79
Standard C	31
Standard C++	1,024
Microsoft C++	2,048
GNU C++	Unlimited
Java	Unlimited

Table 4.8 – Maximum length of identifiers

While Java and **GNU C++** have no restriction on the number of characters in the name of an identifier, you should be reasonable in the number of characters you use.

Coming up with a meaningful name for identifiers is an important task for making your code readable, so give it some thought. Now, let us look at data that, once assigned a value, cannot be changed.

Constants

A constant can be any data type declared with the `final` keyword. It must have a value assigned when declared, for example, `final double TAX_RATE = 0.05;`.

If you declare a field in a class as `final`, then you may also assign its value in the class constructor. Once a value is assigned to a constant, however, it cannot be changed.

The naming rules for constants are the same as identifiers. What differs are the conventions. Constants are nouns written in uppercase. You can use the underscore to separate words in the identifier, as shown in `TAX_RATE` in the previous example.

Operators

Java supports the common set of operators found in almost every language, as shown in the following table:

Action	Operator	Assignment
Addition	x = x + y	x += y
Subtraction	x = x – y	x -= y
Multiplication	x = x * y	x *= y
Division	x = x / y	x /= y
Modulus (remainder)	x = x % y	x %= y
Increment	++x or x++	N/A
Decrement	--x or x--	N/A

Table 4.9 – Basic math operators

Java follows the standard rules of precedence, except for the increment and decrement operators. If the operator is on the left-hand side of the variable, then Java conducts the operation before any other. This, technically, gives it the highest precedence. Placed on the right-hand side, it has the lowest precedence, and Java performs it after all other operations are complete. We will review the logical operators when we look at logical operations in the next chapter.

String operator

String does not have a numeric value, and you cannot use a String object in a calculation. As you would expect, int numberOfDogs = 23; does not mean the same as String numberOfDogs = "23 ";.

You cannot use a String variable in an arithmetic expression. If the characters in String match what is allowable for a number, then you must convert String into a numeric variable first and then use it in an arithmetic expression.

However, the plus (+) operator is permitted along with String. When used with String, it means to concatenate or join multiple String values into one, as shown here:

```
String animal = "moose";
String favoriteAnimal
    = "My favorite animal is a " + animal;
```

You can also use concatenation to combine String with any of the eight primitive types. This will automatically convert the primitive to String, as shown here:

```
int numberOfMoose = 36;
String message = "There are " + numberOfMoose
    + " in the park. "
```

The `String` message will contain the following:

There are 36 moose in the park.

You cannot assign a numeric type directly to `String`. You must concatenate it to `String` or use the `String.valueOf` method; it is a simpler approach. Concatenating a primitive to any string, such as the empty String shown here, will work:

```
String piecesOfSilver = "" + 23;
```

Alternatively, you can use the following:

```
String piecesOfSilver = String.valueOf(23);
```

Where you see the literal values, you can also use a primitive, as shown here:

```
int silverCoins = 23;
String piecesOfSilver = String.valueOf(silverCoins);
```

There is just one operator you can use with String, which is for concatenation rather than a mathematical function.

Casting

Casting provides the ability to cast or convert one data type to another. When coding with primitives, there are two types of casting – implicit and explicit. First, look at this chart, which shows the primitives by order of their range of values:

Largest					Smallest
double	float	long	int	short	byte

Table 4.10 – Relative number of bytes between types

What this means is that when assigning a primitive of one type on the chart to a primitive higher on the chart, Java will perform an implicit cast, as shown here:

```
int apples = 23;
double fruit = apples;
```

This is a lossless conversion. You may try the assignment in the other direction, as follows:

```
double apples = 34.6;
int fruit = apples;
```

You will get the following error:

```
incompatible types: possible lossy conversion from double to
int
```

If you need to convert from a data type larger than the data type of the destination variable, you must do an explicit cast. For example, when converting from a floating point to an integer, Java will cut the fractional component. No rounding; it just disappears, as shown here:

```
double apples = 34.6;


int fruit = (int)apples;
```

With casting, there is no error, but the value in fruit will be 34.

If the value on the right-hand side exceeds the range of the type you are casting to, when it is an integer type, then, like an **overflow**, discussed in the next section, it will wrap around, as shown here:

```
double apples = 67000.6;
short fruit = (short)fruit;
```

The value in fruit will be 1464.

The syntax for casting Java places the type to cast in parentheses. Python, C, and C++ are cast by placing the value in parentheses, while C# follows the same model as Java.

In *Table 4.10*, you did not see the char type. Its sole purpose in the language is to represent a UTF-8 code that Java will render as a character on the screen. You can assign a character to a char variable, or you can assign an integer. If the integer you assign is outside the allowable range, then you will need to cast, and an overflow wrap will occur.

Here are some examples of declaring a char variable:

```
char letterA1 = 'A';
char letterA2 = 66;
char letterA3 = (char)65601;
```

All three of these will become the letter A.

We have seen that promotion occurs implicitly, moving from one data type to another data type higher on the chart. To move in the other direction, you must cast.

Overflow and underflow

An overflow and an **underflow** can occur when working with floating-point types. Only an overflow can occur with integer and char types. Here is how Java behaves in these situations.

Integer overflow

An overflow occurs when a value is outside the range of allowable values. For floating-point values, an overflow results in the special value infinity, and either plus or minus is the result.

Overflow with integers results in a wraparound. For example, in the following code snippet, we are assigning a value to a short data type. This value is 1 greater than the allowable upper range of short:

```
short testValue = (short)32768;
System.out.printf("testValue = %d", testValue);
```

The following is the output of this code fragment:

```
testValue = -32768
```

Floating-point overflow

Unlike the integer types, floating-point types do not wrap when overflow occurs. Instead, Java assigns the special value Infinity.

Using a double as an example, we can use the Double class, discussed in the next section, as the static constant that contains the maximum allowed value for a double. When we assign a value to a double that exceeds the maximum allowable value:

```
double testValue = Double.MAX_VALUE + Double.MAX_VALUE;
System.out.printf("testValue = %f", testValue);
```

The following will be the output for this:

```
testValue = Infinity
```

The nature of the floating point is such that a minor increase over the maximum value does not result in an overflow. As shown in the previous example, a significant increase over the maximum value will generate Infinity. Take the following example:

```
double testValue = Double.MAX_VALUE + 1.0;
double testValue = Double.MAX_VALUE ;
```

Both expressions return the same answer:

```
testValue = 179769313486231570000000000000000000000000000000000
00000000000000000000000000000000000000000000000000000000000000000
00000000000000000000000000000000000000000000000000000000000000000
00000000000000000000000000000000000000000000000000000000000000000
00000000000000000000000000000000000000000000000000000000000000000
0000000000000000.000000
```

Floating-point underflow

Underflow occurs when the floating-point value cannot represent exceedingly small fractions. Like overflow, this condition does not occur immediately after a value falls below the minimum value for the floating point. A meaningful change that lowers a value below the minimum will trigger an underflow. In this example, we begin by assigning the smallest allowed value to a double. When we divide it by 2, it becomes even smaller but now is smaller than the minimum value:

```
double testValue = Double.MIN_VALUE;
System.out.printf("testValue = %2.16e", testValue);
```

When run, the result is as follows:

```
testValue = 4.9000000000000000e-324
```

This is the minimum value of a double. Say we try to assign an even smaller value by dividing the minimum value by 2, as we have done here:

```
double testValue = Double.MIN_VALUE / 2;
```

Then, the result will be the following:

```
testValue = 0.0000000000000000e+00
```

This is the value that the code returns when a float or double underflows.

You should always be wary of overflow and, in the case of floating point, underflow. Now, let us look at a family of classes that provide class support for the primitive data types.

Wrapper classes

Java, like most languages, has an array data type – you can have an array of integers, Booleans, or primitives. You can have an array of objects, such as the String. Every element in the array must be the same type. Many of the object-oriented capabilities in Java require the use of objects and not primitives. For example, Java has a library of data structures called collections that provides greater functionality than a basic array. These collections can only store or collect objects. You cannot have a collection of int, double, or any of the other primitives.

A **wrapper class** is the solution Java creators developed for situations where an object must be used rather than a primitive. Every primitive data type has an equivalent wrapper class, which can be used when a primitive is not permitted. Just like with the String class, you do not need to use new to create a wrapper object. Wrappers have methods to convert from String to primitives. They also contain information about the primitives. We already had a peek at this when I used Double.MIN_VALUE and Double.MAX_VALUE.

Here is a table of all the primitives and their matching wrapper classes. Aside from static variables, these wrappers also have static methods to convert to and from String:

Primitive data type	Wrapper class
byte	Byte
short	Short
int	Integer
long	Long
float	Float
double	Double
boolean	Boolean
char	Character

Table 4.11 – Primitives and wrappers

These classes, like String, do not have to be explicitly instantiated. This means that you could write the following:

```
Integer number = Integer.valueOf(12);
```

But you can also just assign the integer value, as shown here:

```
Integer number = 12;
```

Java refers to this as **autoboxing**. A second feature, called **unboxing**, permits the reading of a wrapper class as if it were a primitive.

Here we see an object, number, assigned to a primitive:

```
int value = number;
```

This now allows us to use objects as primitives. Every wrapper also contains several useful methods we will explore as we delve more into the language.

The math library

Earlier in this chapter, we examined the operators available for working with primitive data types. There are many operations you may wish to perform that do not have a matching symbol, such as raising a value by a power. There are languages that use the caret (^) or double asterisk (**) to denote raising to a power. In Python, you would write the following:

```
value = 5.0**2.0;
```

The result will be 25. Java does not have a symbol for this operation. Instead, we must use a method that belongs to the math class, as shown here:

```
double value = Math.pow(5.0, 2.0);
```

We have already seen this in the program that calculated compound interest; take a look again:

```
var result = principal * Math.pow(
                1 + annualInterestRate / compoundPerTimeUnit,
            time * compoundPerTimeUnit);
```

The math library has an extensive selection of math operations. See the link in the *Further reading* section to learn about all the choices available to you.

Summary

The heart of every program you will write is the data that your program operates on. In this chapter, we have learned about the eight primitive types. There is byte, particularly useful if you are writing software to interact with other devices. short, int, and long are useful when what you need to describe has no fractions. When there are fractions, however, you can use floating-point types – float and double. The char type is the building block for strings. If you want to keep track of what is true or false, you should use the boolean type.

As you move forward into Java, always keep in mind the available data types. Just as important is to understand what will happen if a value is out of range.

Having identified the types, we moved to identify variables with meaningful names. We discussed how we assign data to these variables, and how we use them was an important part of this chapter.

We left the primitives briefly to look at classes that are closely associated with primitives. There was `String` – home to characters that make up text that we can read. The wrapper classes provided runtime information about their matching primitives and are freely interchangeable with primitives, when you need an object rather than a primitive.

Coming up, we will look at classes focusing on access control, packages, and how we construct a class.

Further reading

- *The IEEE 754 Format*: `http://mathcenter.oxford.emory.edu/site/cs170/ieee754/`

- *Demystifying Floating Point Precision*: `https://blog.demofox.org/2017/11/21/floating-point-precision/`

- *Class Math*: `https://docs.oracle.com/en/java/javase/17/docs/api/java.base/java/lang/Math.html`

5

Language Fundamentals – Classes

An **object-oriented** (**OO**) program is based on the design of structures called classes that are used as the blueprint for objects. An object is the implementation of a class. This means that the first step in coding in **OO programming** (**OOP**) is to create classes. This chapter will examine how the features of OOP are implemented in Java. We begin by looking at how we define variables in a class followed by how we control access to members of a class and the class itself. From here, we will look at the class structure Java provides us for creating or working with classes and objects.

In this chapter, you will learn about the following:

- Class fields
- Understanding access control
- Understanding classes

By the end of this chapter, you will be able to define classes, instantiate them into objects, and interact with other classes. Let's begin by looking at access control. Before we begin, let's look at the two categories of variables that you can declare in a class.

Technical requirements

Here are the tools required to run the examples in this chapter:

- Java 17 installed
- Text editor
- Maven 3.8.6 or a newer version installed

You can find the code from this chapter in the GitHub repository at https://github.com/PacktPublishing/Transitioning-to-Java/tree/chapter05.

Class fields

Variables declared in a class and not in a method call are referred to as fields. They fall into two categories:

- Instance variables
- Class variables

Instance variables are unique to every instance of the class. They can be primitives or references. If we had an instance variable of type `double` in a class and we created 100 instances of the object, we would have 100 doubles.

In Java, you can have a variable in a class that is shared by all objects created from the class. In other words, every object has a unique set of instance variables, but all share the **class variables**. This is accomplished by designating the variable as static. There is only one memory allocation for a static variable. In our 100 instances of an object, there is just one double if you declare the double as static.

Another characteristic of a static or class variable is that assuming it has public access control, you can access it without instantiating the object. For example, consider the following code block:

```
public class TestBed {
    public static String bob;
```

In this fragment, we can access the `class` variable `bob` by simply writing `TestBed.bob`. We don't have to instantiate the object. If we do instantiate it, we can use the reference, though this is rare.

The fact that class variables are shared by all objects makes them an ideal tool for objects created from the same class to communicate with each other. If one object changes the value of a class variable, then even another object of the same class can see the updated value.

Before we move on to the next section on access control, let's clarify one more term. We refer to all variables declared in a class as **fields**. This includes both class and instance variables.

Understanding access control

One significant and invaluable feature of OOP is **access control**. If you have already worked with an OO language, then you may be familiar with this concept; if not, let me explain what access control means.

Access control in Java concerns the visibility of classes, fields, and methods to other classes. You must have sufficient access to create objects and access fields and methods in a class.

Access control, in other languages, may imply a security mechanism that can ensure that a request for access to a method–for example–is coming from an authenticated user. This is not what Java does; in Java, it is about how objects can interact with each other.

Let's look at the options for visibility; the first will be Java packages.

Packages

The first piece of the access control puzzle is the Java package and its corresponding `import` statement. In *Chapter 3*, *The Maven Build Tool*, we learned about Java packages and how they are simply a folder that contains Java code. Code in one package cannot access code in another package unless you give access by including an `import` statement. The class that imports can access code in what it imports. Without an `import` class, the `statement` classes in different packages cannot interact with each other.

Keep in mind that the interaction is one way. For example, class A has an `import` statement for class B. Class A can call or send messages to code in class B but not vice versa. You could add an `import` statement to class B for class A, and then they can each send a message to the other.

Using packages for access control is simplistic; it is a simple binary setting. You can either see an object of a class that you are importing or you cannot see the class at all if you did not import it. All classes in the same package have an implicit import for the other classes in the same package, and, in this case, there is no need to import fellow package classes explicitly.

Now, we are ready to look at the four access control specifiers available to us.

The public specifier

The **public specifier** in Java is the same as in C++ and C#. It defines how users can access fields and methods of a class. With the appropriate `import` statement, if required, any other object can create objects of that public class.

You can access a public class's fields from any object that has a reference to a second object. We should always keep class fields private so that we cannot directly assign values to the field. To interact with a private variable, you will need methods in the same class that will be your proxy to read or write to the private variable. When writing to the field, you will be able to validate the new value before it is assigned to the field.

You can call class methods that are public from any other object that has a reference to the object containing the public method. We refer to the public methods of a class as its interface. We will look at interfaces in the next chapter.

The private specifier

The **private specifier** in Java is the same as in C++ and C#–you use this specifier to define access control for fields and methods.

As already mentioned, class fields should always be private. You will need to validate the data that you want to store in a private field. A common way to do this is with a mutator, commonly called a setter method. Here, you can add validation and reject invalid data by throwing an exception.

Methods designated as private can only be called upon by other methods in the same class. Using private methods allows us to break down complex tasks into smaller units. As they are private, they cannot be called upon from other objects, and this ensures that all the necessary steps of a complex task will be carried out in the correct order.

Java permits you to define a new class within another class. This is the only situation where a class may be private. A private non-inner class could not be instantiated. You cannot use a private inner class outside the class you declare it in. You can instantiate a private inner class in the class that declared it. As already mentioned, instance variables should always be private while methods may be any of the four access designations.

The protected specifier

The **protected specifier** in Java is similar to those in C++ and C#–you use this specifier to define access control for fields and methods. Protected, also referred to as `protected/package`, is only used when you are employing **inheritance**. In a non-inheritance situation, protected behaves the same as package access. In C++ and C#, the concept of package does not exist, so in a non-inheritance situation, these languages treat protected as private.

Inheritance, as we will see in the next chapter, is an arrangement between two classes where one class is a **superclass** and the other is a **subclass**. A subclass can access all the public members of its superclass, including the instance variable designated as protected in the superclass. Other objects with references to the superclass not involved with inheritance see protected members as private or–if these objects are defined in the same package–as package.

Methods and class variables that you designate as protected also possess package access, as described in the next section. You cannot have a protected class.

The package specifier

This final access specifier, package, has no equivalent in C++ or C#, although there is some similarity to the friend concept in these languages. It defines the visibility of classes, fields, and methods in other objects that are defined in the same package. There is no designator such as public, private, or protected. When not explicitly using a specifier on a class, field, or method, then the implicit access control is package.

An object that has a reference to another object defined in a different package or folder sees protected as private. The objects of two different classes in the same package can access protected elements as if they were public.

> **Note**
>
> One last point before we move on—use protected and package specifiers sparingly. They exist for situations where the interaction between objects can be speeded up. The problem is that they expose fields and methods to objects that should not have access to them. I recommend that you only ever use public and private when you start designing and coding a program. Only if you can demonstrate that the system performance is suffering by only using public or private components should you consider protected and package.

One last point to reiterate—access control does not exist inside a class. This means that a public method can call a `private` method in the same class. Every method has access to every field no matter its access control designation.

Now, we are ready to look at how classes work. We will first look at the classes from the last chapter's program, and then we will create an updated version of the program.

Understanding classes

Java, as with other OO languages, uses a syntax that revolves around the source code structure called a **class**. But first, what is a class? The theorists who introduced the concept of objects envisioned a class as a custom data type. Think of the primitive integer type—it has a range of allowable values and a pre-defined set of operations such as addition, subtraction, and the other usual operators. Imagine a class as a custom primitive in which you decide which operations, in the form of methods, your type will perform. One goal of OOP is to focus on problem-solving by developing custom data types that combine data and actions as opposed to the structured programming approach where data and actions are separate.

This means that you develop a class by first listing all the fields of the class, either primitives or references to other classes. Next come the methods that conduct useful tasks that make use of these fields. These variables are visible to every method in a class regardless of the access control level.

Classes are not executable code, with one exception we will see shortly. Instead, a class is a blueprint that must be instantiated or created at runtime using the new keyword. When your program begins execution, the **Java virtual machine** (**JVM**) stores the class definition in a region of memory called the **class-method region**.

The new operator conducts two tasks:

- Allocates sufficient memory for all the instance variables in a class blueprint. We refer to this region of memory as the **heap**. Memory for class or static variables is allocated when the program begins and is found in the `Class-Method` region.

- Assigns the address of the allocated regions of memory to the appropriate reference variable. A reference variable is always just 4 bytes long. Accessing objects through a reference variable is called indirect addressing.

There is a third region of memory that comes into play called the stack. The JVM stores all local variables–variables declared in methods–in the stack data structure as required. The stack is a dynamic structure that can allocate space for variables and then deallocate them by just moving a pointer. If you are interested in memory management, see the *Further reading* section at the end of this chapter for more information.

Before we go any further, we need to understand the two methods that the JVM can invoke whenever we create an object.

constructor and finalize methods

There are two special methods found in most OO languages related to memory management. The `constructor` method runs as the last step in the creation of an object whereas the `destructor` method, called `finalize` in Java, runs as the first step when an object goes out of scope. How this happens in Java is different than in other languages such as C++.

finalize

There can be only one `finalize` method in a class. You cannot overload it. Do not use it–Java 9 deprecated it. If you are a C++ developer, you may mistakenly believe that `finalize` is the Java equivalent of a C++ destructor. However, that's not true–here's why.

In C++, the `delete` operator runs the `destructor` method first. Once the `destructor` method runs, the `delete` operator releases the memory used by the object, which can now be reallocated. This works because when you execute the `delete` operator on a valid pointer to an object, the actions in the `destructor` function execute immediately.

In Java, there is no `delete` operator; instead, the JVM monitors all references to objects. When an object reference goes out of scope, the JVM enters it on a list of references that the JVM will release for you. However, for efficiency and performance, the JVM does not immediately release memory. Instead, it puts it off for as long as possible. This is due to the time necessary to release memory, which can impact the current programs running in the JVM. We call the release of memory **garbage collection**.

The JVM runs the `finalize` method just before garbage collection. This garbage collection could be occurring every few minutes or less, but that's very unlikely. Given a great deal of RAM (my system has 32 GB), this can mean that garbage collection could happen every few hours or even days. Therefore, `finalize` is non-deterministic. It may want to effect changes or send messages to parts of your program for reasons no longer valid, or that part of the program has already been garbage collected.

For this and other reasons, the Java architects made the decision to deprecate `finalize`. So, do not use it. If you want a method called before an object goes out of scope, then you will have to explicitly call a method you created.

With the `finalize` method discussed and then assigned to the rubbish or deprecated bin, let's look at the constructor.

Constructor

The purpose of the constructor is to conduct any actions necessary as the last step in object creation. Typically, you use the constructor to initialize class instance variables. We have seen that we can initialize these variables right at the point of declaration in the class. Sometimes, the initialization requires more steps than just assigning a value, and this is where the constructor is invaluable.

You create an object by invoking the new operator. The new operator conducts tasks that result in the creation of an object. I am simplifying how this works, but it conveys what you need to know:

1. Memory is allocated in the heap region of memory for all instance variables, along with other required structures that the JVM requires. The address of this memory is what the this reference captures. Subsequently, every call to non-static methods in this class now has a first parameter, which is the invisible this parameter.

    Let's code the following:

    ```
    public class MyClass() {
        public void doSomething(int value) { … }

            . . .

    }
    ```

    Then, we'll instantiate it, like so:

    ```
    var testClass = new MyClass();
    ```

    The non-static methods become the following after being compiled:

    ```
        public void doSomething(MyClass this, int value) { … }
    ```

    You can never write the code this way. MyClass this is implied and so may not be written.

    Now, we'll call this method:

    ```
    testClass.doSomething(42);
    ```

    It then becomes the following:

    ```
    testClass.doSomething(testClass, 42);
    ```

2. Once memory is allocated, along with other housekeeping chores, the JVM calls the appropriate constructor method. It does not have a return type as it does not have a variable to return a result to. You must name them with the same name as the class they belong to.

Constructors fall into two categories: default and non-default. A **default constructor**, of which there can only be one, has no parameters. A **non-default constructor** is one with parameters. It is subject to overloading, so there can be more than one non-default constructor if the parameter types are different.

Java provides the capability of having one constructor call another constructor. The first constructor called is determined by the rules of overloading. The called constructor can then call upon one other constructor. This call must be the first line of code. Watch out for a possible `recursive constructor invocation` error where constructor A calls constructor B and constructor B calls constructor A.

Revising the compound interest program

We are now ready to review our compound interest program and apply what we have just covered to the classes in this project.

For this, let's look in more depth at the `CompoundInterest04` program that we discussed in *Chapter 2, Code, Compile, and Execute.*

We will begin by declaring the package in the `CompoundInterestCalculator04.java` file. Packages, the folders we place our source code into, allow us to manage our code by functionality. The only time you might not want to use a package is if you are creating a single-file source-code or Linux shebang application.

Here is the package declaration. The file will be in a folder named `business`, which is in the `com/kenfogel/compoundinterest04` folder:

```
package com.kenfogel.compoundinterest04.business;
```

This program will be using the `NumberFormat` class. This class is part of Java's standard library, and we know this because the first name in its package designation is `java`. To make use of this class, we must import it into our file, like so:

```
import java.text.NumberFormat;
```

The first line of the class declaration shows us the following:

- This class is public and so can be instantiated in any other class that declares a reference to it.
- The public class name, `CompoundInterestCalculator04`, must also be the name of the file. You can have more than one class structure in a file but only one of them may be public.

Here is the first line of the declaration:

```
public class CompoundInterestCalculator04 {
```

Here are the fields in the class:

```
        private final double principal = 100.0;
        private final double annualInterestRate = 0.05;
```

```
    private final double compoundPerTimeUnit = 12.0;
    private final double time = 5.0; //
```

We are declaring four instance variables of type double. The access control designation of private means that you cannot access these variables from any other class that may have a reference to this class. The final modifier defines these variables as immutable. In a class, access control does not apply but modifiers do. You must initialize a final variable where you declared it or in the constructor.

Next, we are declaring references to objects that we will use in the code:

```
    private final NumberFormat currencyFormat;
    private final NumberFormat percentFormat;
```

We are declaring two instances of a NumberFormat class. You can see from the variable names that we planned each one for a different format. These are final, meaning we must initialize them with a value and they cannot be instantiated a second time. Rather than instantiate the NumberFormat references in their declaration, we can also instantiate them in a constructor, and that is what we will do.

The following method is the constructor:

```
    public CompoundInterestCalculator04() {
        currencyFormat =
            NumberFormat.getCurrencyInstance();
        percentFormat = NumberFormat.getPercentInstance();
        percentFormat.setMinimumFractionDigits(0);
        percentFormat.setMaximumFractionDigits(5);
    }
```

The constructor is easy to recognize as it must have the same name as the class. It does not return a value as the JVM calls the constructor as part of the new operation. There is nothing to assign a result from a return statement to. This is a default constructor as there are no arguments inside the parentheses. A class may have only one default constructor but may overload the constructor with constructors that take as arguments different data types.

The task that this constructor conducts is the initialization and configuration of the NumberFormat objects. Rather than just using the new operator, this class is instantiated by using a factory method. Factory methods conduct additional tasks before invoking new. Also, take note that we call the methods through the class name and not through an object name, very much like the Math library methods. This tells us that getCurrencyInstance and getPercentInstance are static methods that are available. We will discuss static methods later in this section.

Next up is the method we want to call after the object is instantiated:

```
    public void perform() {
```

The name `perform` is just a name of my choosing. All that is important to remember is that, except for the constructor, all methods should be verbs. Keep in mind that class and variable identifiers should be nouns.

The first line of the method calls upon the `calculateCompoundInterest` method to perform the calculation and stores the result in the `result` variable:

```
var result = calculateCompoundInterest();
```

This next line displays the result formatted appropriately:

```
System.out.printf(
    "If you deposit %s in a savings account "
    + "that pays %s annual interest compounded "
    + "monthly%nyou will have after %1.0f "
    + "years %s%n",
    currencyFormat.format(principal),
    percentFormat.format(annualInterestRate),
    time, currencyFormat.format(result));
}
```

The plus symbols in the code mean concatenation. As the string is quite long, it has been broken up into multiple strings joined by the plus operator.

Here, we see the method that carries out the calculation of the answer:

```
private double calculateCompoundInterest() {
    var result = principal *
      Math.pow(1 + annualInterestRate /
      compoundPerTimeUnit, time * compoundPerTimeUnit);
    return result;
  }
}
```

Using the instance variables, calculate the result to a variable named `result` that the method returns to whoever called it. It is a private method, so only other methods in this class can see it.

The second class from the compound interest example holds just the `main` method.

Again, let's declare the package this file is in:

```
package com.kenfogel.compoundinterest04.app;
```

This program will be using the CompoundInterestCalculator04 class that we have written. As with all imports, we are referencing a class that we wrote in a package/folder we created:

```
import com.kenfogel.compoundinterest04.business
                              .CompoundInterestCalculator04;
```

Here is the first line that declares the class:

```
public class CompoundInterest04 {
```

It shows us the following:

- This class is public and so can be instantiated in any other class that declares a reference to it.
- The class name, CompoundInterest04, must also be the name of the file.

You can have more than one class structure in a file but only one of them may be public.

Every Java program must have a main method. This is where the JVM begins the execution of your program:

```
public static void main(String[] args) {
```

The main method is a static method. What sets static methods apart from non-static ones is that we can call these static methods without instantiating the object, such as what we did with the static methods for NumberFormat. The Math library works the same way. To use the pow function, we just write Math.pow. We do not need to instantiate the Math object first.

Here, we instantiate the CompoundInterestCalculator04 class into an object with a reference named banker:

```
var banker = new CompoundInterestCalculator04();
```

We end main with a call to the perform method in CompoundInterestCalculator04 class:

```
banker.perform();
    }
}
```

We have now reviewed how the CompoundInterest04 program is constructed and how we make use of access control and packages.

Class organization based on functionality

The compound interest program we have been using from the start of this book conducts its specific task correctly. However, the problem with this approach is that the program is a dead end–not that being a dead end is necessarily bad. Sometimes, you just want a *one-off* program that can determine the answer to a specific problem.

But what if we were to write a complex program? In the real world, meaning coding in your job, you will seldom be writing programs such as our compound interest calculator. Imagine you wanted to create a more thorough banking system. In this system, it will be necessary to gather input from the user rather than hardcoding it in the program's source code. Taking it further, you may want to store the input and the result in an external storage such as a database. You may also want to generate reports based on the stored data.

Let's now reorganize the `CompoundInterest04` program, now renamed `CompoundInterest05`, based on functionality.

The data class

The first step is to design a class that just holds the data. There will be no domain methods, such as for calculating, storing in a database, or interacting with the end user for input and output methods. We are creating a new data type that we can use in classes that perform other actions. This type of class follows a pattern first described as a **JavaBean**. Java introduced it as a reusable software component.

What we are creating is not a pure JavaBean, but a variant. I frequently refer to this type of class as a simple box of variables. Let's look at one for our compound interest problem.

We begin with the `package` statement. We will use the package name in any class that needs to use this by importing the name, like so:

```
package com.kenfogel.compoundinterest05.data;
```

Here is the standard public class declaration:

```
public class CompoundInterestData {
```

Here are the four variables required to perform the calculation:

```
        private final double principal;
        private final double annualInterestRate;
        private final double compoundPerTimeUnit;
        private final double time;
```

We declare the four instance variables here. They are final so that once assigned a value, it becomes immutable. We expect these to come from the program's user rather than hardcoding the values as we have done up until now. This means that every new calculation will need a new CompoundInterestData object.

This last variable is where we plan to store the result of the calculation:

```
private double result;
```

As there are no actions in this class, we cannot determine when this value will be set, so it cannot be final.

This is the constructor:

```
public CompoundInterestData(double principal,
        double annualInterestRate,
        double compoundPerTimeUnit,
        double time) {
    this.principal = principal;
    this.annualInterestRate = annualInterestRate;
    this.compoundPerTimeUnit = compoundPerTimeUnit;
    this.time = time;
}
```

It has four required arguments when this class is instantiated. Once assigned to the class variables, you cannot change the value they hold. Take note of the this keyword. As we used the same name for the instance variables and the arguments of the method, we use this to designate an instance variable. Remember that this is the address of the instance variables in the class. There is not a this reference for class or static variables as there is ever only one of them per class. Without using this, you are referring to the method's argument of the same name. The result instance variable is not one of the arguments as you will calculate its value later in the program.

The next four methods are getters for the four class instance variables:

```
public double getPrincipal() {
    return principal;
}

public double getAnnualInterestRate() {
    return annualInterestRate;
}

public double getCompoundPerTimeUnit() {
```

```
        return compoundPerTimeUnit;
    }


    public double getTime() {
        return time;
    }
```

The JavaBean specification that this class is based on requires that all instance variables be private. The spec goes on to define setter and getter methods. As these first four are immutable by virtue of being final, you can only have a getter. You will be providing the initial value when you create an object using new.

These last two methods are special because the result variable is not final:

```
    public double getResult() {
        return result;
    }


    public void setResult(double result) {
        this.result = result;
    }
}
```

We determine the value after this class is instantiated. We assign values to the four input values in the constructor. Frameworks such as the Java Persistence API or Jakarta expect this getter and setter syntax.

The business class

Now, let's write the calculation class. Its sole purpose will be to calculate the result and store it in the bean. We are importing the data class we just created:

```
package com.kenfogel.compoundinterest05.business;


import com.kenfogel.compoundinterest05.data.
                                    CompoundInterestData05;


public class CompoundInterestCalculator05 {
```

This class has only one method and does not have any instance variables. As all the values are in CompoundInterestData05, we retrieve the values with a call to the property's getter method. We end by assigning the result to the bean's result variable by calling the only setter:

```
    public void calculateCompoundInterest(
                CompoundInterestData05 value) {
        var result = value.getPrincipal() *
                Math.pow(1 + value.getAnnualInterestRate() /
                value.getCompoundPerTimeUnit(),
                value.getTime() *
                value.getCompoundPerTimeUnit());
        value.setResult(result);
    }
}
```

The user interface class

The last component is the user interface where we can ask the user for the four pieces of information required to perform the calculation. This is where we will create this object:

```
package com.kenfogel.compoundinterest05.ui;
```

After the `package` statement, we have imports for the classes and libraries we will use. We have a new import, and that is for the `Scanner` library class. Objects of the `Scanner` class allow us to gather end-user input, such as from the keyboard, in a console application:

```
import com.kenfogel.compoundinterest05.business.
                    CompoundInterestCalculator05;
import com.kenfogel.compoundinterest05.data.
                    CompoundInterestData05;
import java.text.NumberFormat;
import java.util.Scanner;


public class CompoundInterestUI05 {

    private CompoundInterestData05 inputData;
    private final CompoundInterestCalculator05 calculator;
    private final Scanner sc;
    private final NumberFormat currencyFormat;
    private final NumberFormat percentFormat;
```

In Java, there are no rules covering the order of methods or the locations of fields, unlike in C and C++ where the order can be significant. My personal style is, as you see in these examples, to place the instance variables first and the constructors right after. My advice on this point is to have the team you work with all agree to the coding style. This will make reading each other's code far easier.

Here is the constructor that instantiates the `NumberFormat` objects and the `Scanner` class. You must provide the `Scanner` class's constructor with the source of input. It could be from a file on disk, but for this program, it is coming from the keyboard. We call the object that interacts with the keyboard's `System.in`:

```java
public CompoundInterestUI05() {
    currencyFormat =
            NumberFormat.getCurrencyInstance();
    percentFormat = NumberFormat.getPercentInstance();
    percentFormat.setMinimumFractionDigits(0);
    percentFormat.setMaximumFractionDigits(5);


    sc = new Scanner(System.in);
    calculator = new CompoundInterestCalculator05();
}
```

Next up is the entry point for this user interface class. This will be the only public method in this class. I use the `do` prefix as it ensures that the name is a verb or action. The four values we must request from the user exist as local or method variables. We assign the result of user input to each one. We instantiate the data object with the four local variables as parameters in this expression. The new operator copies the values from the local variables to the instance variables of the `CompoundInterestData05` object through the constructor. We then call upon `calculateCompoundInterest` in our `Calculator` class to calculate the result. The last step is to display the result:

```java
public void doUserInterface() {
    doUserInstructions();
    var principal = doPrincipalInput();
    var annualInterestRate = doAnnualInterestRate();
    var compoundPerTimeUnit = doCompoundPerTimeUnit();
    var time = doTimeInput();
    inputData = new CompoundInterestData05(
            principal,
            annualInterestRate,
            compoundPerTimeUnit,
            time);
```

```
calculator.calculateCompoundInterest(inputData);
displayTheResults();
}
```

This version of the CompoundInterest program follows the classic pattern of input, process, and output. The first method we now encounter is output:

```
private void displayTheResults() {
    System.out.printf(
      "If you deposit %s in a savings account that pays"
      + " %s annual interest compounded monthly%n"
      + "you will have after %1.0f years %s%n",
      currencyFormat.format(inputData.getPrincipal()),
        percentFormat.format(
            inputData.getAnnualInterestRate()),
            inputData.getTime(),
            currencyFormat.format(
                inputData.getResult())));
}
```

This next method is part of the input process. It could provide additional instructions on the screen to the end user, but I have kept it simple here:

```
private void doUserInstructions() {
    System.out.printf(
            "Compound Interest Calculator%n%n");
}
```

Now, we come to the user input. Each input will display a prompt and wait for input. Once you enter the string, the nextDouble method attempts to convert it to the appropriate type–in this case, double:

```
private double doPrincipalInput() {
    System.out.printf("Enter the principal: ");
    var value = sc.nextDouble();
    return value;
}
private double doAnnualInterestRate() {
    System.out.printf("Enter the interest rate: ");
    var value = sc.nextDouble();
```

```
            return value;
        }
    private double doCompoundPerTimeUnit() {
        System.out.printf("Enter periods per year: ");
        var value = sc.nextDouble();
        return value;
    }
    private double doTimeInput() {
        System.out.printf("Enter the years: ");
        var value = sc.nextDouble();
        return value;
    }
}
```

But wait–there is something terribly wrong with the four input methods. They are identical except for the String prompt. As a teacher, I described repeated code as an invitation to failure. If we decided to make a change, such as switching from double to float, we must remember to make this change four times in eight separate places. The potential for inadvertently neglecting one of the changes is too high. Let's turn this from four methods to one.

It's simple–just make the prompt an argument to a single input method, like so:

```
    private double doUserInput(String prompt) {
        System.out.printf(prompt);
        var value = sc.nextDouble();
        return value;
    }
```

Now, we can use the doUserInput method for all four user inputs:

```
    public void doUserInterface() {
        doUserInstructions();
        var principal =
            doUserInput("Enter the principal: ");
        var annualInterestRate =
            doUserInput("Enter the interest rate: ");
        var compoundPerTimeUnit =
            doUserInput("Enter periods per year: ");
        var time = doUserInput("Enter the years: ");
```

```
                inputData = new CompoundInterestData05(
                        principal,
                        annualInterestRate,
                        compoundPerTimeUnit,
                        time);
                calculator.calculateCompoundInterest(inputData);
                displayTheResults();
        }
```

All user input is a `String` object; you cannot enter pure numbers, Booleans, or characters. The `Scanner` class is responsible for converting from a string to the destination type, as expressed by the next methods.

In our example, they are all doubles. What happens if we enter the bob string instead of a number? Java will throw an exception. This is an error condition. When we look at loops, we will learn how to create user-proof input, and when we look at GUI programming, we will learn about other ways we can manage user input. In all situations, all input arrives–as already mentioned–as strings.

The last class is the app class. We commonly use the designation app to define a package that contains the class that holds the `main` method. This is a convention, and you can freely change this:

```
package com.kenfogel.compoundinterest05.app;
import com.kenfogel.compoundinterest05.ui
                                .CompoundInterestUI05;


public class CompoundInterest05 {
    public static void main(String[] args) {
        var calculator = new CompoundInterestUI05();
        calculator.doUserInterface();
    }
}
```

When we run this new version, this will be the output:

```
Welcome to the Compound Interest Calculator

Enter the principal: 5000
Enter the interest rate: 0.05
Enter periods per year: 12
Enter the years: 5
```

```
If you deposit $5,000.00 in a savings account that pays 5%
annual interest compounded monthly
you will have after 5 years $6,416.79
```

The program requests from the user the four values it requires, and then using those values, it calculates the result and displays it.

Summary

In this chapter, we have explored the basic components of a class. Once we reviewed how we assembled the `CompoundInterest04` example, we broke the program apart and created classes to hold the data, display a user interface, and calculate the result. We also learned about the constructor and the deprecated `finalize` methods. We gained insight into what `new` does and how the JVM manages a program's memory.

The second version, `CompoundInterest05`, presents how a program is organized professionally based on functionality. It kept separate the data, the user interface, and the action, usually called the business. To gather user input, we had our first look at the Java library `Scanner` class. You should now have a good understanding of how a Java class is organized and how you can control access to members of the class.

In the next chapter, we will look more closely at the methods that carry out the actions of a class and how we manage the relationships between classes.

Further reading

- *Stack Memory and Heap Space in Java*: `https://www.baeldung.com/java-stack-heap`

6

Methods, Interfaces, Records, and Their Relationships

In Java, how we define and organize our code is the cornerstone of the language. In this chapter, we will begin by examining the role that a method plays in Java. From here, we will examine the relationships afforded by inheritance and interface. The immutable record class is next up. Polymorphism, the ability to use objects in a hierarchy of classes, as it applies to inheritance and interface, is covered next. We will finish the chapter by looking at the relationships between objects and how they can call upon methods in other objects.

We will learn about the following topics in this chapter:

- Understanding methods
- Understanding inheritance
- Understanding the class interface
- Understanding the record class
- Understanding polymorphism
- Understanding composition in classes

By understanding all the components and relationships available in Java, you'll be able to read or write Java code.

Technical requirements

Here are the tools required to run the examples in this chapter:

- Java 17 installed
- A text editor
- Maven 3.8.6 or a newer version installed

The sample code for this chapter is available at `https://github.com/PacktPublishing/Transitioning-to-Java/tree/chapter06`.

Understanding methods

Now, I must admit that I have been coding since 1980, and since then, the term we use to describe discrete blocks of code has changed. When I started coding in 1980 using **BASIC**, an unstructured language, I quickly learned to break my code up into subroutines. From BASIC, I moved on to **Pascal**, where there is a formal designation for blocks of code. These were **functions** for blocks that returned a result and procedures for blocks that did not return a result. Next up was C, followed by C++. These languages name their blocks of code as functions, as they all, except for the constructor, must return a value. Moving on to Java, these blocks are called methods. Let us examine the components of a method.

When creating a method, you will need to consider some or even all of these components:

- Access control designation
- Static or non-static designation and the `this` reference
- Override permission
- Override required
- Return type
- Method name
- Parameter variables
- Annotations
- Exception handling
- Thread setting
- Generic parameters

Let's discuss these, one by one.

Access control designation

In the previous chapter, I presented the concepts of access control, so let us examine how these concepts can be applied to a method, which may be private, public, protected, or package:

- A private method, using the `private` designation, can only be accessed by non-static methods in the same class:

```
private void doSomething() { … }
```

- Public methods can be accessed by non-static methods in the same class. They can also be accessed from any other object in the program that has a reference variable to the object that contains the public method:

  ```
  public void doSomething() { ... }
  ```

- Protected methods can be accessed as if they were public from any subclass that inherits them. We will discuss inheritance later in this chapter. A protected member of a class also has package access. Without inheritance and with objects in a different package, protected behaves the same as private:

  ```
  protected void doSomething() { ... }
  ```

- Package methods in one object can be accessed by another object in the same package that has a reference to it as if it were public. Objects in other packages, if they have the appropriate reference, consider package methods private and cannot access them:

  ```
  void doSomething() { .. } // Package
  ```

Static or non-static designation and the this reference

A method is non-static by default. This means that when the method is invoked, there is always an undeclared first parameter. If other parameters exist, then this undeclared parameter always comes first.

The undeclared parameter is a reference to the object invoking the method. Its type is the class, and the identifier is this.

To understand this, let us look at some code fragments. Here is a partial class with just one method. Since the field and the method parameter have the same name, we will use the this reference to differentiate between them:

```
public class StaticTest01 {
    private int value;
    public void nonStaticMethod(int value) {
        this.value = value;
    }
}
```

The this reference comes from the compiler adding it as the first parameter to the non-static method. Let us look at how the method appears after the compiler has added the this reference:

```
public class StaticTest01 {
    private int value;
    public void nonStaticMethod(StaticTest01 this, int value){
```

```
        this.value = value;
    }

    ...

}
```

We instantiate the object, as shown here:

```
    var sTest = new StaticTest01();
```

Then, we call upon its method, as shown here:

```
    sTest.nonStaticMethod(42);
```

Then, the method is converted by the compiler as shown:

```
    sTest.nonStaticMethod(sTest, 42);
```

The this reference allows just one block of code for methods regardless of how many instances you create of a class.

You can use the this reference in the body of the method, but you cannot declare it. As shown earlier, the most common use of this is to distinguish between a field identifier and a local method variable with the same identifier name.

A static method, the one which you have designated by adding the static keyword to the method, does not have the this reference. This means that you can call this method without instantiating the object it belongs to first. It also means that it cannot call non-static methods in the same class you declared it in, nor can it access any non-static fields.

Override permission – final

Inheritance is the creation of new classes derived from an existing class. We will look at this topic in detail later in this chapter. This means that a method in the derived class or subclass with the same name and the same type and number of parameters overrides the same method in the parent or superclass.

The nonStaticMethod method can be overridden if StaticTest is a superclass with an inheritance relationship, but we can prevent this if we add the final keyword to the method declaration. Now, we can no longer override nonStaticMethod, as shown in the following code block:

```
public class StaticTest {
    private int value;
    public final void nonStaticMethod(int value) {
        this.value = value;
    }
}
```

We can also use the `final` keyword when we declare the class. This will block this class from being used as a superclass. In other words, you cannot inherit from a `final` class:

```
public final class StaticTest {
```

Now, we cannot extend this class. This also means that all the methods in the class are effectively final, so it is not necessary to use the keyword in method names.

Override required – abstract

While using or not using the `final` keyword determines whether a method can be overridden when inheritance is involved, we can also require that the method is overridden. We have just discussed how we can control whether you can override a method or not. Another choice is to make subclassing mandatory so that you *must* override the method. We accomplish this by defining an `abstract` class and declaring abstract methods that do not have any code, as shown in the following code block:

```
public abstract class ForcedOverrideTest {
    private int value;
    public abstract void nonStaticMethod(int value);
    }
}
```

A class with one or more abstract methods cannot be instantiated. You can only use it as a superclass in inheritance. Therefore, we must also designate the class as abstract.

Return type

In Java, as with C, all methods return a result. When we declare a method, except for a constructor, we must show a return type. A return type may be either a primitive variable or a reference variable.

There is one special return type, and that is `void`. Using `void` means that this method does not have a `return` value and that there cannot be a `return` statement that returns a value in the method:

- This method returns nothing:

```
public void doSomething() { … } //
```

- This method returns a primitive:

```
public double doSomething() { … }
```

- This method returns a reference:

```
public String doSomething() { … }
```

Depending on the logic you are employing in a method, there may be times when you wish to return or break out from a method early. In this case, when void is the return type, you may use return by itself. If the method is not void, then the early returns must also include the appropriate value.

Method name

The rules and conventions for naming a method are identical to that of naming a variable we discussed in *Chapter 4*, *Language Fundamentals – Data Types and Variables*. One difference is that while variables are nouns, based on naming conventions, methods are expected to be verbs.

Parameter variables

After the method name comes an opening and closing parenthesis; if left empty, this method does not receive any data. As already pointed out, for non-static methods, there is an undeclared this reference added in the first position of the parameter list. A non-static method that we write without any parameters gets the reference. This effectively means that all non-static methods have at least one parameter.

Parameters can be any primitive type or a class reference type. A method can have no more than 255 parameters, although you are likely doing something wrong if you have anywhere near 255 parameters.

When you call a method, all that is significant is the type of values you pass. The identifier is not significant. If you have a parameter of the int type named bob, you can pass any named int to the method.

Java does not have a way to set a default value for a parameter.

Annotations

An annotation is a hint. It can modify the behavior of your code at compile time or just be informative without any effect. An annotation for a method must appear before the method and any of its components.

An annotation is a phrase that follows the same naming conventions as a class and begins with the at symbol (@). It may have parameters within parentheses that follow the annotation, but these are not variable declarations as in method parameters. They can be anything.

Frameworks, such as **Spring** and **Jakarta**, to name just two, use numerous annotations. For example, if we were writing a Servlet class, a class instantiated and run by a web application server, we would annotate it as follows:

```
@WebServlet(name = "AddToEmailList",
                    urlPatterns = {"/AddToEmailList"})
public class AddToEmailListServlet extends HttpServlet{...}
```

In this example, the annotation defines this class as a `WebServlet` class called `AddToEmailList` rather than using the class name `AddToEmailListServlet`.

You can use annotations anywhere, not just for the web.

Exception handling – throws

When we look at the syntax of Java code in the next chapter, we will come across situations where you can either predict or expect errors when the program is running. For example, you are trying to open a connection to a database server, but the operation fails. The code that tries to make the connection will throw an exception. An exception is just an object that contains the details of the error.

In Java, the exception can either be checked or unchecked. A checked exception expects there to be code to deal with the exception; if you do not handle the exception, it is an error. An unchecked exception does not need to be handled and may or may not end your program.

You can also decide that when an exception occurs, it will not be handled in the method in which the exception occurred. Instead, you want to pass the exception back to the method's caller. You do this with a `throws` clause. Here is a fragment of a method that connects to a database to retrieve a list of items in the database.

In the following code block, we have a method that opens a connection to a database:

```
Connection connection;
private void openConnection(){
    connection =
        DriverManager.getConnection(url, user, password);
}
```

If the connection fails, an exception of the `SQLException` class type, which is a checked exception, will occur. This code will result in a compile-time error because you are not handling the checked exception. You can decide to defer handling the exception in the method that called this one. To do this, we add a `throws` clause to the method as shown in the following code block:

```
Connection connection;
private void openConnection() throws SQLException{
    connection =
        DriverManager.getConnection(url, user, password);
}
```

The compiler will now verify that this exception will be handled in the caller of this method.

Why not handle the exception when it occurs? We have already seen our `CompoundInterest` program broken into a user interface class and a business calculation class. Imagine that you decided to handle the error where it happened by requesting new credentials from the user such as a username or password. How would you do this? Do you ask in the console, a GUI, or a web application?

After asking the user for their database credentials in the user interface class, you pass this information to the `openConnection` method in the business class. If something goes wrong, we throw the exception and return to the calling method in the user interface class. The database class does not need to know what the user interface is. It just returns a result or an exception. This business class is now usable regardless of the user interface. We call this **Separation of Concerns**; we will explore this topic in *Chapter 10, Implementing Software Principles and Design Patterns in Java*.

Thread setting

Code blocks that run concurrently use threads, as we will see in *Chapter 9, Using Threads in Java*. In some cases, you want a block of code in a thread to finish before another thread can run the same block. One way we can force this to happen is to indicate that a method must run to completion before another thread can execute the same block. This is the role of the `synchronized` keyword, as shown in the following code block. This keyword is not foolproof but is a part of how we write thread-safe code:

```java
synchronized public int add(int num1, int num2) {
    return num1 + num2;
}
```

Generic parameters

The concept of generics is present in many languages. This means that you can write code in which the data type of variables is determined by a generic parameter. We will look more closely into this in *Chapter 7, Java Syntax and Exceptions*, but for now, you should recognize the generic parameter syntax.

In the following code block, we have declared a method to return part of a larger list. A list is like an array, and we will look at it in detail when we discuss **collections** in *Chapter 8, Arrays, Collections, Generics, Functions, and Streams*. While we do not know what the generic parameter, T, will be, this code ensures that the result returned is the same type as the parameters passed to the method:

```java
public <T> List<T>getSubList(List<T> a, int b, int c) {
    return a.subList(b, c);
}
```

This code fragment will return a list of objects of a type to be determined at compile time. The first `<T>`, the generic parameter, informs the compiler that T is a type parameter that will be determined by whatever code calls this method. In this example, it shows that the given `List` can be of any type of object, and this method will return a sub-list of the same type.

Whatever language you are transitioning from, I don't suspect you ever considered using at least 11 different concepts to declare a method. There is no requirement that you use all 11 when declaring a method. That we have as many as 11 pieces available is one of the aspects of Java that makes it the ideal language for a wide range of applications.

We will look at the basic syntax of the Java language as used in methods in the next chapter. Next up, let us understand the relationships that classes and objects can have with each other.

Understanding inheritance

What do you do when you have a class that does almost everything you need to get done? What do you do when you have methods in a class that don't quite do what you need to get done? How do you deal with these two issues if you do not have access to the source code of the class and its methods? What if you do have access to the source code, but other parts of your program expect the original unchanged code? The answer is **inheritance**.

Inheritance is described as a relationship between two classes in which one class is called the **superclass**. This class contains the fields and methods to handle a specific task. This superclass is a generalization, and sometimes you need to enhance this kind of class. In this case, rather than rewriting the original class, you can create a new class called the **subclass** that inherits or specializes the superclass by overriding methods in the superclass or adding additional methods and fields. The subclass is now made up of the public and protected methods of the superclass as well as what it has added. A subclass may also be called a derived or child class in other languages.

Imagine that you must write a program that manages bank accounts. All accounts, such as chequing and savings, have similar tasks they must perform. You must be able to deposit money and withdraw money. At the end of the defined period, such as monthly, there are tasks that must be carried out that are similar but not quite the same for our two account types. Here is where inheritance can be useful.

In this program, we define a superclass with common elements. We then create two subclasses that inherit the public and protected members of the superclass. This is commonly described as an **is-a** relationship, meaning that the subclass is a type of superclass. We will see how we use this relationship when we look at **polymorphism**.

Here is a **Unified Modeling Language** (UML) diagram. This diagram style is useful for planning what classes you will need to solve a problem and what the relationship between these classes will be. We begin by simply creating a box for each class and then joining the boxes with lines. The line ending describes the relationship. The hollow triangle signifies inheritance.

In the following figure, we see that `BankAccount` will be the superclass while `SavingsAccount` and `ChequingAccount` are subclasses:

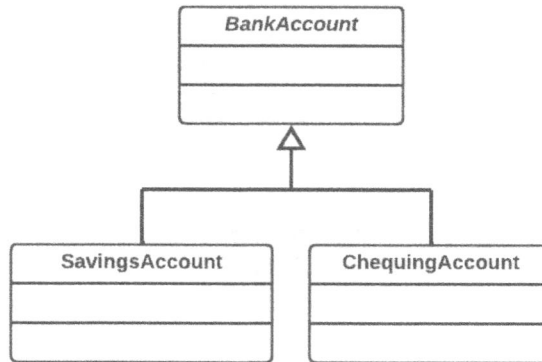

Figure 6.1 – The BankAccount inheritance hierarchy

In this example, we use inheritance to share data and functionality between the superclass and the subclasses. There will never be an object of the `BankAccount` type, only objects of the `SavingsAccount` and `ChequingAccount` types. This will mean that `BankAccount` will be an abstract class. An abstract class, denoted by the class name being in italics, cannot be instantiated into an object; it can only be used as a superclass, so `BankAccount` will hold the data elements. These elements must be accessible to the subclasses but are private to any other classes in the system. To define access control in a UML diagram, we use the following prefixes:

- An octothorpe (#) means the method or field is protected
- The plus sign (+) means the class, method, or field is public
- The minus sign (-) means the class, method, or field is private
- The absence of a prefix means the method or field is package

Here, we are showing that all the fields in `BankAccount` are protected and available to the two subclasses:

BankAccount

balanceAtStartOfMonth: double
currentBalance: double
totalOfCurrentMonthDeposits: double
countOfCurrentMonthDeposits: int
totalOfCurrentMonthWithdrawals: double
countOfCurrentMonthWithdrawals: int
annualinterestRate: double
serviceChargeCurrentMonth: double
accountStatus: boolean

SavingsAccount

ChequingAccount

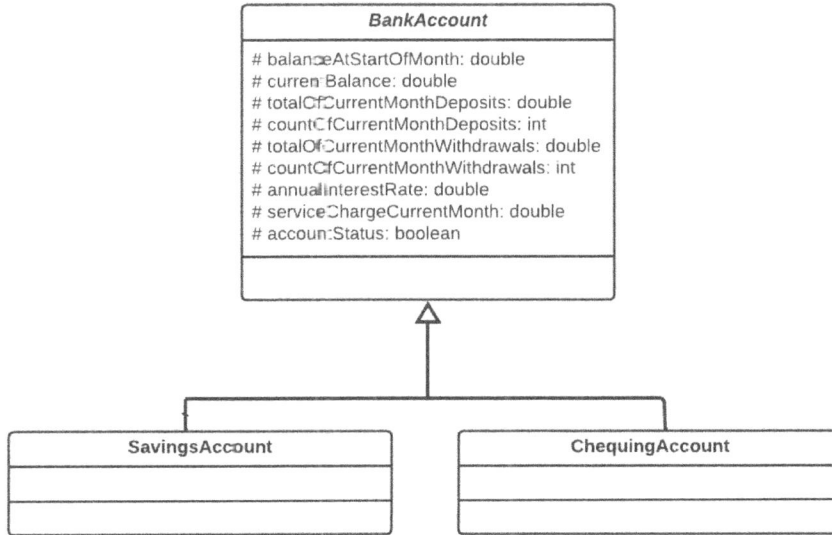

Figure 6.2 – The BankAccount class fields

The last part of this inheritance design is to determine what methods will be required. Here we can see the methods in the superclass and subclasses:

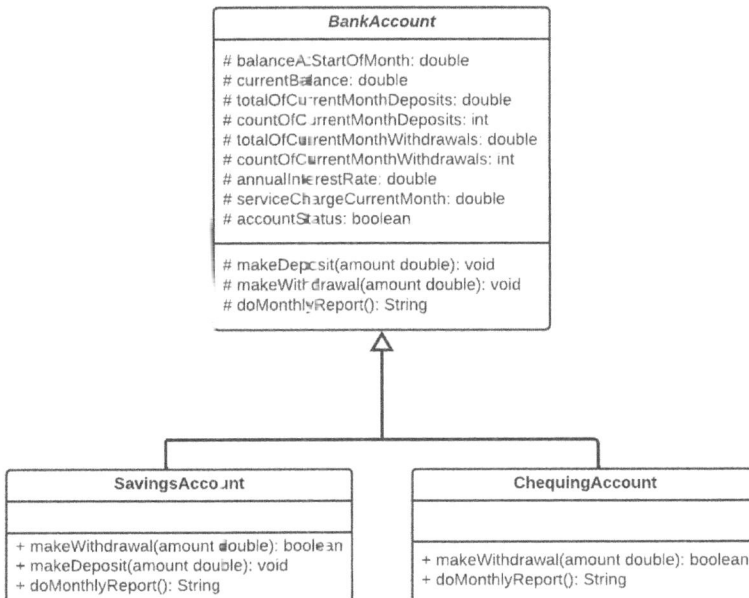

BankAccount

balanceAtStartOfMonth: double
currentBalance: double
totalOfCurrentMonthDeposits: double
countOfCurrentMonthDeposits: int
totalOfCurrentMonthWithdrawals: double
countOfCurrentMonthWithdrawals: int
annualinterestRate: double
serviceChargeCurrentMonth: double
accountStatus: boolean

makeDeposit(amount double): void
makeWithdrawal(amount double): void
doMonthlyReport(): String

SavingsAccount

+ makeWithdrawal(amount double): boolean
+ makeDeposit(amount double): void
+ doMonthlyReport(): String

ChequingAccount

+ makeWithdrawal(amount double): boolean
+ doMonthlyReport(): String

Figure 6.3 – Methods and their access control

In the case of BankAccount, the three tasks—deposit, withdrawal, and report—each have a method that performs the actions common to both account types. These methods will be called upon by the subclasses.

SavingsAccount, by its nature, must override each of the BankAccount methods. To override means to have a method with the same name and the same parameters. The return type can be different but must be a subclass of the original return type. The methods in the subclass can call upon the overridden method in the superclass by prefixing the method call with the super reference, such as super.makeDeposit(amt).

ChequingAccount only overrides two of the superclass methods. It does not override makeDeposit, as how the superclass handles this is sufficient.

Inheritance is a one-way street. The subclass can call upon public and protected members of the superclass, but the superclass is unaware of the subclass.

The inheritance model that only overrides superclass methods in the subclass and does not add any additional public methods or variables in the subclass is referred to as **pure inheritance**. The inheritance model that adds additional methods and variables to the subclass and may or may not override any superclass methods is described as an **is-like-a** relationship and is called an **extension**.

We code inheritance of both types using the extends keyword. Assuming we have defined the BankAccount superclass, then we code inheritance as follows:

- public class SavingsAccount extends BankAccount { … }
- public class ChequingAccount extends BankAccount { … }

We will need to examine the code in the subclass to determine which approach, pure inheritance or extension, is taken.

When instantiated, each of these objects will have its own dataset from its superclass and, if the inheritance is an extension, then its own fields.

In the following example, the BankAccount class will be declared as abstract. This means that you cannot instantiate BankAccount as follows:

```
var account = new BankAcount();
```

This will be flagged as an error by the compiler because BankAccount is abstract. If we had any abstract methods in BankAccount, we would be required to override them in SavingsAccount and ChequingAccount.

> **Note**
>
> The C++ language supports **multiple inheritances**. This means that a subclass can have more than one superclass. The designers of Java chose to only support single inheritance. You can only extend a class with a single superclass.

If you do not want your class to be available for inheritance, you can add the `final` designation, as shown here:

```
public final AClass { … }
```

If we now attempt to create an inherited or derived class, we see the following:

```
public class AnotherClass extends AClass { … }
```

An error will be declared when we try and compile this code.

The superclass of all objects, the Object class

In Java, all classes extend a special class called **Object**. While we cannot extend more than one class, `Object` is always available. Why does the `Object` class exist? The `Object` class defines methods to support threading and object management in any class. This means that every class has these methods and may choose to override any of them or use them as is. There are three of these methods that are frequently overridden in a class. The first of these is as follows:

```
public boolean equals(Object obj)
```

The default implementation inherited from `Object` compares the address of the object that invokes `equals` to the object passed as a parameter. This means it can only be `true` if two references contain the same address in memory to an object. In most cases, you will likely want to compare the values in an object's fields, and therefore you will frequently want to override it. For example, see the following:

```
public class Stuff {
    private int x;
    public Stuff(int y){
        x = y;
    }
    @Override
    public boolean equals(Object obj) {
        if (this == obj) {
            return true;
        }
```

```
        if (obj == null) {
            return false;
        }
        if (getClass() != obj.getClass()) {
            return false;
        }
        final Stuff other = (Stuff) obj;
        return this.x == other.x;
    }
```

The equals method performs four tests that must all be passed for one object to be equal to another. These are the following:

- The this reference and obj contain the same memory address
- The comparative obj is not null
- Both references are for the same class type
- Finally, it compares the value of the fields

The next commonly overridden method is this:

```
public int hashCode()
```

When comparing multiple fields in two objects, the necessary code can be quite time-consuming. If you need to compare two strings, it will be necessary to compare them one character at a time. You will need to compare each one for a class with many fields. There is an optimization, and that is **hashCode**.

A hash is an integer computed from the fields in an object. Computing hash values is, in most cases, faster than comparing fields one at a time. The default value from this method, if it is not overridden, is the address in memory at which the object resides. What you will want to do is calculate a hash value based on the object's fields.

Here is an example of a method that overrides the Object class's hashCode method with its own:

```
    @Override
    public int hashCode() {
        int hash = 5;
        hash = 79 * hash + this.x;
        return hash;
    }
```

The value generated is not a unique value. It is possible that two objects with different values would generate the same hash code. What we do know for certain is that if the hash codes of two objects are not the same, then the objects, based on the fields, are not equal. If the hash code is the same for two objects, then to be certain that they are equal, you must now use the `equals` method. This will, in almost all cases, speed up the process of comparing objects because the slower `equals` method is only called when the hash codes are the same.

One other application of `hashCode` is in data structures—structures that store data as a pair of values called the key and value, where the value is an object. The value that `hashCode` returns is an integer value that is used as the key. As the integer is processed faster than any of the other primitives, these data structures perform more efficiently as compared to structures that can use any type as the key.

The last of the three commonly overridden methods is this:

```
public String toString()
```

The implementation of this method in the `Object` class returns the object's address as returned by the `hashCode` method and the name of the class. Overriding it to return the values of the fields as a String can be far more useful. By overloading it, you can inspect the state of an object, as shown here:

```
@Override
public String toString() {
    return "Stuff{" + "x=" + x + '}';
}
```

Let us now look at an alternative to inheritance in Java, a **class interface**.

Understanding the class interface

In a programming language that has access control features, public methods are called the class **interface**. These are the methods that can be called from any other object in a system that has a reference to the object it wants to use. In Java, we can create a contract that will require any class that implements the contract, called an interface, to implement all the methods listed in the interface as public methods.

Here is an interface for reading and writing to a relational database:

```
public interface GamingDAO {

    // Create
    int create(Gamer gamer) throws SQLException;
    int create(Games games) throws SQLException;
```

```
    // Read
    List<Gamer> findAll() throws SQLException;
    Gamer findID(int id) throws SQLException;

    // Update
    int update(Gamer gamer) throws SQLException;
    int update(Games games) throws SQLException;

    // Delete
    int deleteGamer(int ID) throws SQLException;
    int deleteGames(int ID) throws SQLException;
}
```

In this code block for an interface class, each method is declared as an abstract method, as it ends with a semicolon and does not have a code block. They are all public by default as well – hence, the lack of the public keyword.

We use an interface when we declare the class. The implementing class must now have public methods as described in the interface. The following are the first few lines of a class that implements GamingDAO. I have not included the implementation of these methods:

```
public class GamingDAOImpl implements GamingDAO {
    @Override
    public List<Gamer> findAll() throws SQLException {…}

    @Override
    public int create(Gamer gamer) throws SQLException {…}

    @Override
    public Gamer findID(int id) throws SQLException {…}
```

Each method has the @Override annotation. This is an informative annotation and is optional in this case. Your code will be more easily understood by other developers if you use this annotation. You are also informing the compiler to watch out for any changes in the superclass methods and if they are found, the compiler will complain. This is the original application of a Java interface.

Java 8 and then Java 9 modified what can be contained in an interface. The changes to the original definition of an interface class are as follows:

- Having the default interface method. This is a public method implemented in the interface rather than in the implementing class.

- Having `private` methods implemented in the interface class. These `private` methods can only be called by a `default` method, or they can be called by other `private` methods declared in the interface. Therefore, as in the following code block, I can call `write4()` from `write2()`.

- Having `static` methods implemented in the interface class. As with all `static` methods, it does not have a `this` reference, so it cannot call upon other methods in the interface or class. It must be a public method, so the `public` keyword is not required.

Here is an example of these three types of methods that can exist in an interface class:

```
public interface Pencil {
    void write1(); // Standard interface method
    default void write2() {
        System.out.printf("default%n");
        write4();
    }
    static void write3() {
        System.out.printf("static%n");
    }
    private void write4() {
        System.out.printf("private%n");
    }
    public void perform(); // Standard interface method
}
```

If we implement this interface, then the only method that we are under contract to implement is `write1()`:

```
public class WritingTool implements Pencil {
    @Override
    public void write1() {
        System.out.printf("standard interface");
    }

    @Override
    public void perform() {
        write1();
        write2();
        Pencil.write3();
```

```
    }

    public void write5() {
        System.out.printf("Method only in WritingTool");
    }

    public static void main(String[] args) {
        Pencil tool = new WritingTool();
        tool.perform();
        tool.write5();
    }
}
```

Notice that in the `main` method, the object we are creating, `WritingTool`, has its reference assigned to an object of the `Pencil` type. You cannot write `new Pencil()`, but you use `Pencil` as the reference type when you create `WritingTool`. This will restrict your use of the code in the `WritingTool` class to just the overridden methods. The last call in the `main` method, `tool.write5()`, will generate a compiler error because the `write5` method is not part of the `Pencil` interface.

The implementing class, in this case, `WritingTool`, can have additional methods of any access control designation not listed in the interface.

It is considered a best practice to define the public methods of a class in an interface class. It is not necessary to list every public method and a class may have multiple interfaces so that what you can do with a class can be restricted to a specific set of tasks.

Abstract class versus interface

In an abstract superclass, you must implement every abstract method in the subclass. This might lead you to believe that an abstract class with every method abstract is the same as an interface. However, as mentioned here, there are significant differences between the two.

An abstract class may have the following:

- Abstract method declarations
- Class and instance variables (fields)
- A constructor
- An additional non-abstract method of any access control

Whereas, an interface may have these:

- Abstract method declarations
- Default, private, or static methods

There is one other significant difference – as Java only supports a single inheritance model, you can only extend a single superclass. On the other hand, you can implement multiple interfaces for a single class, as shown here:

```
public class SingleInheritance extends ASuperClass { ... }
public class MultiInterface implements IFace1,IFace2 {...}
```

You can also have a subclass with a single superclass and one or more interfaces, as shown here:

```
public class MultiClass extends ASuperClass implements
                                    IFace1,IFace2 {...}
```

However, one issue to keep an eye out for when using multiple interfaces is that it is a compile-time error, should there be the same abstract method in more than one interface. You may have an abstract method in a superclass identical to an abstract method in an interface.

Both interfaces and abstract classes can define what a class must implement. The fact that a class can only inherit from a single class makes an interface class more useful, as a class may have more than one interface.

Sealed classes and interfaces

By default, any class can be the superclass for any other class, just as any interface can be used with any class. However, you can restrict this behavior by using a **sealed class** or **sealed interface**. This means you can list the names of classes that may extend or implement it, as shown in the following snippet:

```
public sealed class SealedClass permits SubClass{ }
```

Here, we have declared a class that can only be used as a superclass to a class whose name is SubClass.

Now that we have permitted SubClass to extend the Sealed class, we can write this:

```
public final class SubClass extends SealedClass { }
```

Note that this subclass must be defined as final. This way, it is not possible to have SubClass as a superclass for some other class. Here is a similar syntax for a sealed interface:

```
public sealed interface SealedInterface permits SubClass{ }
public final class SubClass implements SealedInterface { }
```

We have one more structure to examine, and that is `record`.

Understanding the record class

A **record class** simplifies the creation of a class that can only have immutable fields. **Immutability** means that the fields of a class are all final and must have a value assigned to them when the `record` object is instantiated.

At its simplest, a record only needs the fields listed when the `record` class is declared, as shown here:

```
public record Employee(String name, double salary) { }
```

When we instantiate this record, we must provide the values for name and salary:

```
var worker = new Employee("Bob", 43233.54);
```

All records have a default canonical constructor that expects a value for every field in the record. In a regular class, you would need to write a canonical constructor method to assign the values to the fields. You may add a compact constructor to a record that permits you to examine the value each field was assigned.

Here is a compact constructor. Notice that it does not have a parameter list, as it can only have a parameter for every field declared in the first line of the class:

```
public record Employee(String name, double salary) {
    public Employee {
        if (salary < 50000) {
            // code if this is true
        }
    }
}
```

To access the value of a field, you can use the identifier as a method as shown here:

```
var aSalary = worker.salary();
```

You cannot change the value of a field in a record – you can only read its value. As a class, a record also extends `Object`. Furthermore, `record` provides default overrides for the `equals`, `hashCode`, and `toString` methods based on the fields.

A record can implement an interface and so can be in the list of a sealed interface. As a record is implicitly final, it cannot extend any other classes, so it cannot be sealed for a class or classes.

Now, let us examine the concept of polymorphism and see how it allows us to reuse code.

Understanding polymorphism

Polymorphism is one of the defining features of object-oriented languages. Java has expanded the concept by including interfaces. Let us begin with a simple hierarchy of classes as shown here:

```
public class SuperClass {

    protected int count;

    public void setCount(int count) {
        this.count = count;
    }

    public void displayCount() {
        System.out.printf('SuperClass count = %d%n", count);
    }
}
```

In this trivial class, we have a public method to assign a value to count and a second method that displays the value in count along with the name of the class. In the following code block, we have a class that uses SuperClass:

```
public class Polymorphism {
    private void show(SuperClass sc) {
        sc.setCount(42);
        sc.displayCount();
    }

    public void perform() {
        var superClass = new SuperClass();
        show(superClass);
    }

    public static void main(String[] args) {
        new Polymorphism().perform();
    }
}
```

When we run this, the result is as we expected:

```
SuperClass count = 42
```

Now, let us create a subclass from `SuperClass`:

```java
public class SubClass extends SuperClass {
    @Override
    public void displayCount() {
        System.out.printf("SubClass count = %d%n", count);
    }
}
```

This subclass only overrides the `displayCount` method. Classes are about creating new data types. When we create a class that is a subclass, we refer to the class type by the superclass name. In other words, we can state that `SubClass` is of the `SuperClass` type. If we added a `subclass` variable to `SubClass`, then that class is also of the `SuperClass` type. Now let us change the `Polymorphism` class to use a `SubClass` object:

```java
public void perform() {
    var subClass = new SubClass();
    show(subClass);
}

private void show(SuperClass sc) {
    sc.setCount(42);
    sc.displayCount();
}
```

While the `show` method is unchanged and still expects an object of the `SuperClass` type, we are creating an object of `SubClass` in `perform`, and then we call upon `show`, passing a reference to the subclass. As `SubClass` is of the `SuperClass` type, polymorphism permits any subclass of `SuperClass` to be passed to the `show` method. When we call `sc.setCount`, the runtime determines that it must use the `superclass` count method because there is no public method with that name in the subclass. When it calls upon `sc.displayCount`, it must decide whether to use the method that belongs to `SuperClass` that it is expecting or the method that belongs to the `SubClass` type that is passed.

Polymorphism means that methods overridden in a subclass are called in favor of the superclass version even when the declared type of what is being passed is the `SuperClass` type. The result when running the code this time is the following:

```
SubClass count = 42
```

Classes that share the same interface are also subject to polymorphism. Here is a simple interface that only requires the method:

```
public interface Interface {
    void displayCount();
}
```

Any class that implements this interface must override the abstract displayCount method. Now, let us create a class that implements this interface:

```
public class Implementation implements Interface {
    protected int count;

    public void setCount(int count) {
        this.count = count;
    }
    @Override
    public void displayCount() {
        System.out.printf("Implement count = %d%n", count);
    }
}
```

Now, let us use polymorphism with the interface:

```
public class PolyInterfaceExample {
    private void show(Interface face) {
        face.displayCount();
    }
    public void perform() {
        var implement = new Implementation();
        implement.setCount(42);
        show(implement);
    }
    public static void main(String[] args) {
        new PolyInterfaceExample().perform();
    }
}
```

In this class, we are passing the `Implementation` object to a method expecting the interface. When this runs, it produces the following output:

```
Implementation count = 42
```

Any class that implements the interface named `Interface` can be passed to any method that declares a parameter using the interface rather than a class.

Polymorphism is a powerful tool that allows you to write code that evolves over time but does not require changes to existing code that uses objects that share an interface or inherit from the same superclass.

Understanding composition in classes

When we create an application that employs numerous classes, we must decide how they will interact with each other. In object-oriented programming terminology, a method in one class that calls a method in another class is called messaging. Despite this, most developers describe this as calling a method, as I do. How objects send these messages or call methods in other objects is what **composition** is about.

There are two ways for objects to be connected – **association** and **aggregation**. Let's discuss these connections.

Association

In association, the object reference we need to call or message a method is created outside the calling object. Let's begin with a class that we will want to call:

```java
public class Receiver {
    public void displayName(String name) {
        System.out.printf("%s%n", name);
    }
}
```

This is a trivial method that has a method to display a String that is passed to it. Keep in mind that we are interested in the concepts here and not the task the class is responsible for. Now, let us create a class that will want to call or message `Receiver`:

```java
public class Association {

    private final Receiver receiveString;

    public Association(Receiver receiveString) {
        this.receiveString = receiveString;
```

```
    }
    public void sendMessage() {
        receiveString.displayName("Bob");
    }
}
```

In this class, we declare a reference to the `Receiver` class. We do not plan for it to be changed when it is assigned an initial, so we designate it as `final`. We do not create an instance of this object using `new`. Rather, we expect this object to be created in another class and we will then pass that reference to the constructor. Put another way, this `Association` class does not own the reference to `Receiver`.

Here is an example of having an object declared in a class and then that object passed to another class:

```
public class Composition {
    public void perform() {
        var receive = new Receiver();
        var associate = new Association(receive);
        associate.sendMessage();
    }

    public static void main(String[] args) {
        new Composition().perform();
    }
}
```

In the `Composition` class's `perform` method, we are instantiating an object of the `Receiver` type. Following this, we instantiate an object of the `Association` type, passing to its constructor a reference to the `Receiver` object we just created.

When the time comes to put an object of the `Association` type out for garbage collection, the `Receiver` object is untouched if it is also used in another object. Put another way, if the `Receiver` object is in scope, meaning visible and valid, in another object, then the `Receiver` object is not garbage-collected. Let's see rewriting `perform` as follows:

```
public void perform() {
    var receive = new Receiver();
    var associate = new Association(receive);
    associate.sendMessage();

    associate = null;
```

```
        associate.sendMessage(); // ERROR

        receive.displayName("Ken");
    }
```

To explicitly put an object out of scope and available for garbage collection, you can assign a special value named null. This will set the address this reference contains to zero, and the object will be made available for garbage collection. If we try to call sendMessage after assigning null, then this will be flagged as an error by the compiler. If we remove the error line, the final line where we call displayName in the Receiver object will work, as the Association class did not own the Receiver object. Ownership is defined as belonging to the object that created it. Association did not create Receiver. If the Association object's receive reference is not in scope anywhere else in the program, it will go out for garbage collection.

Aggregation

Objects that are created in a class belong to or are owned by that class. When we create an object of the owning class, all the other objects in the class will be instantiated in the class. When this class goes out of scope, then everything it owns also goes out of scope.

Here is our program employing aggregation:

```
public class Aggregation {

    private final Receiver receiveString;

    public Aggregation() {
        receiveString = new Receiver();
    }

    public void sendMessage() {
        receiveString.displayName("Bob");
    }
}
```

In this Aggregation class, an object of the Receiver type is created in the constructor. It is not coming from another object as it did with Association. This means that Aggregation owns the Receiver object and when the Aggregation object goes out of scope, so does the Receiver object that it instantiated.

Summary

In this chapter, we have finished examining the building blocks or concepts of organizing code in Java that we began in *Chapter 5, Language Fundamentals – Classes*. We looked at methods and a range of issues that need to be considered when writing a method. From there, we examined inheritance, a way in which we can reuse or share code.

The interface introduced the concept of a contract or list of methods that must be written by any class that implements the interface. The specialized class type called `record` for simplifying the creation of immutable objects was next up.

Inheritance and interfaces support the concept of polymorphism. This permits the creation of methods that expect an instance of a superclass or interface class but receive an instance of any class that either inherits or extends the superclass or implements the interface.

We ended the chapter by looking at how we connect objects to objects. Composition implies that the object is created outside of the object that has a reference. The reference must be passed to the object either through the constructor or another method. Aggregation implies that the object we need to use is created inside the object that wishes to use it.

Next up, we will finally review the syntax of the Java language.

7
Java Syntax and Exceptions

In this chapter, we will begin by looking at the syntax of the Java language. It might seem strange that it took till this chapter to look at syntax. To understand why, I must let you in on a secret: you already know how to code. This is the audience this book is for – you can program but have little or no experience with Java. I have no doubt that you could understand what was happening in every code sample you have seen so far. We will now formalize the Java syntax.

Here are the topics we will cover:

- Understanding coding structures
- Handling exceptions

By the end of this chapter, you will be able to organize Java code into methods and classes. Decision-making and iteration in Java code are presented. When things go wrong, there can be, in many situations, the need to leave the code that caused the error and either carry out additional processing to solve the problem or exit the program. This is the role of exceptions.

Technical requirements

Here are the tools required to run the examples in this chapter:

- Java 17 installed
- Text editor
- Maven 3.8.6 or a newer version installed

You can find the code from this chapter in the GitHub repository at `https://github.com/PacktPublishing/Transitioning-to-Java/tree/chapter07`.

Understanding coding structures

When we write code in any language, we know that it must be organized in very specific ways. You are familiar with this concept from whichever language or languages you already know, so all we must do is examine how they are coded in Java. We begin with code blocks.

Code blocks

Every language has a structure for organizing the lines of code you write, and this is commonly called a **block**. The Python language uses indenting to define a block, and Pascal uses the begin and end keywords. Java uses opening ({) and closing (}) braces, as do C, C++, C#, and JavaScript.

In Java, all classes and methods must have an opening and closing brace. Blocks may be nested, as we will see when we examine iteration and decisions later in this section. Blocks also serve another purpose when it comes to variables. This is called the variable's scope. Let's look at this in practice in an example:

```java
public class Blocks {
2
3      private int classScope;
4      private static int staticVar;
5
6      static {
7          staticVar = 42;
8          System.out.printf(
                "static block staticVar is %d%n",staticVar);
9      }
10
11     public Blocks() {
12         System.out.printf("constructor method block%n");
13     }
14
15     public void perform() {
16         int methodScope;
17
18         if (classScope > 0) {
19             int blockScope = 4;
20         }
21
```

```
22              {
23                  int classScope = 3;
24                  this.classScope = classScope;
25              }
26          }
27      }
```

We'll discuss each line in detail.

Line 1 declares a class named `Blocks`, and an opening brace appears on this line. C/C++ programmers typically place opening braces on their own line, and Java is fine with this. The Java style is to place an opening brace on the same line that names the block.

Line 3 declares an instance field. We know this because it is declared inside the class block and is not static. For every object, there will be a unique `classScope` variable. As this field is in the class block, it is visible to all non-static methods in the class. It is also available to any inner blocks in any method. It only goes out of scope when the object instantiated from this class goes out of scope.

Line 4 declares a static or class variable. This variable will be shared by all instances of the class. It is visible in all blocks in the class. One thing it cannot be is a local method variable. Static variables can only be declared in the class block or scope.

Lines 6 through *9* declare a static block. The code in this block is only ever executed once when the first instance of this object is created. You cannot declare fields in this block. You cannot interact with instance (non-static) variables, but you can interact with class (static) variables. You may also call upon static methods such as `System.out.print` or any static methods in this class. An interesting characteristic of static blocks is that they execute before the constructor. This is why they cannot access non-static variables and methods. These non-static variables are only valid after the constructor executes, not before.

Lines 11 through *13* are just a constructor. If you add a `main` method to this example, you will be able to see that the static block always executes before the constructor.

Lines 15 through *26* are a non-static method block named `perform` that in turn contains two additional blocks. In the `method` block, we have the `methodScope` local variable that is visible and accessible in the method and in any inner blocks. This variable will go out of scope when the execution of the method reaches the closing brace of the `method` block.

Lines 18 through *20* consist of an `if` statement followed by a block that is executed should the `if` statement be `true`. In this block, we have declared a variable named `blockScope`. This variable comes into scope after the opening brace is encountered and the declaration is found. When the block ends, this variable goes out of scope.

Lines 22 through *25* are another block. In here, we are declaring a variable of the same name and type as the class-scoped variable. When this occurs, the block version of a variable hides any variable declared in an outer block, which in this case is the class itself. To access the class block variable, as we have seen when we discussed methods, we use the `this` reference. If you create more blocks in blocks, which is not really a good idea, you can only access the class level variable with `this`, and any variable of the same name and the type in outer blocks becomes inaccessible.

Moving on, let's briefly review the meaning of the terms *statement* and *expression* in Java.

Statements

In Java, any line of code that performs a task and ends in a semicolon is a statement. Here are a few examples:

```
1   int x = 4;
2   printHeader();
3   d = Math.sqrt(aDoubleValue);
```

These are all statements. *Line 1* is a declaration statement where an integer variable is assigned space in memory and a value is assigned to it. *Line 2* is a call to a method. *Line 3* uses the square root method from the `Math` library to compute a result that you are assigning to a variable.

Expressions

An expression in Java is any code that returns a result as part of a statement. The result may be from an assignment, some simple math, or as the return value from another method or Java construct, such as the `switch` expression that we will see shortly.

In the examples in the *Statements* section, we can see that the *line 1* statement includes an expression that assigns a value to a variable. *Line 2* is just a statement as there is no value that is changing. *Line 3* takes the value returned by the call to `Math.sqrt` and assigns it to a variable. When we assign a new value to a variable, we describe this as changing its state. Statements that change the state of a variable do this with an expression.

Operators

Java's family of operators is quite like what is found in C and most other languages derived from C/C++ or modeled on them. The rules of precedence are respected, and expressions inside parentheses are always carried out first. All the standard logical operators exist. As Java does not have pointers, operators that deal with pointers, such as the address of (`&`) and the indirection (`*`), do not exist in this language. There is one group of operators that I do want to highlight.

In C/C++, we express the outcome of combining multiple Boolean expressions in one of two ways—either logical AND or logical OR. They are expressed as a double ampersand (&&) and a double pipe (||). They employ short-circuit evaluation, which means if there's a condition that validates or invalidates the statement in the first comparison, then there is no need to carry out a second comparison. The values on each side of the operator must be Booleans. Here's an example:

```
numberOfTrees > 10 && numberOfSquirrels > 20
```

There is a matching set that can perform the same task but without short-circuit evaluation. These are the single ampersand (&) and the single pipe (|). When working with primitive types, they perform a bitwise operation. For the single ampersand (&), there must be a binary 1 in the same position in each value that becomes a binary 1 in the new value. Otherwise, a binary 0 is placed in the new value. For the single pipe (|), the matching bits must have one of the bits as a binary 1.

There is one more operator in this family worth mentioning and that is the caret (^). This is the XOR operator. As used with the primitive types, the new value takes a binary 1 only if one of the two values being compared has a binary 1 in the same position. Otherwise, the result is 0.

In Java, there is a hierarchy of the numeric primitive types—as listed next—based on their size in memory and the range of allowable values. We saw this earlier in *Chapter 4, Language Fundamentals – Data Types and Variables*:

1. `byte`
2. `char`
3. `short`
4. `int`
5. `long`
6. `float`
7. `double`

Assignment statements have a right-hand side and a left-hand side, such as:

LHS = RHS

Based on this list, you can only have a type on the LHS that has a larger range of values than the type on the RHS. This means you can write something like this:

```
int intValue = 27;
double doubleValue = intValue;
```

This works because the conversion from `int` on the RHS to `double` on the LHS is lossless. In the other direction, as shown next, it will be an error because the fractional part of `double` will be lost:

```
double doubleValue = 23.76;
int intValue = doubleValue;
```

This all leads to the casting operator—the parenthesis plus type. The parenthesis is also used in other ways but when used here, it becomes an operator. To make the previous example work, you can cast `double` to `int`, like so:

```
int intValue = (int) doubleValue;
```

This is a lossy conversion as the fractional component of the `double` value is sliced off, with no rounding. The value that ends up in `intValue` will be `23`.

There is one more operator—the arrow operator (`->`), which we will encounter when we examine the modern switch and functional programming. Let's now move on and examine iteration, commonly called looping.

Iteration

Java provides us with two approaches to iteration. The first, which we will look at now, is the classical looping technique. We will examine using streams to iterate over every member of a collection in the next chapter.

The for loop

Let's begin with the C-style `for` loop. This is a loop where the conditions for iteration are in the first line of the loop inside parentheses:

```
For (int x = 0; x < 10; ++x) {
    System.out.printf("Value of x is %d%n", x);
}
```

The entire `for` loop is considered a block. This means that an x variable is created when the `for` loop is entered, and it goes out of scope when the loop ends. If you need access to x after the loop ends, then declare it before the loop, as shown:

```
int x;
for (x = 0; x < 10; ++x, doMethod()) {
    System.out.printf("Value of x is %d%n", x);
}
```

There are two special statements available in classic loops:

- The `break` statement will end a loop before it finishes iterating

- The `continue` statement ends the current iteration of the loop and moves on to the next iteration

The foreach loop

There is one more style of the `for` loop, called the `foreach` loop. It is predicated on the fact that every element in an array or collection will be processed. We will examine the `foreach` loop when we look at collections in the next chapter.

The while and do/while loops

When a `for` loop is written, the maximum number of iterations is known right away. For our next loops, `while` and `do/while`, the number of iterations cannot be predicted as it will depend on something changing in the body of the loop.

In using `while` and `do/while`, the loop is dependent on something happening inside the loop block, which may change the variable that is being logically examined. Here is an example with an unpredictable ending:

```
var rand = new Random();
int x = rand.nextInt(12);
while (x < 10) {
    x = rand.nextInt(12);
    System.out.printf("x = %d%n", x);
}
```

The first line instantiates the `java.util.Random` object. Next, we instantiate the variable that will be the basis of the logical test and give it a random value. The method call to `rand.nextInt(12)` will return a value with a range of 12 possible integers between 0 to 11 inclusively. This points out that a `while` loop can iterate zero or more times, but it is not possible to predict how many iterations. We express the logical test in the parentheses of the `while` statement. Inside the loop, we must perform some action that alters the state of the x loop variable. There are no restrictions on what you can code in the `while` block.

A variation of the `while` loop is the `do/while` loop. This loop is guaranteed to iterate at least once as the logical test occurs at the end of the loop. You can see it in action here:

```
Var rand = new Random();
int x;
do {
    x = rand.nextInt(12);
    System.out.printf("x = %d%n", x);
} while (x < 10);
```

Notice that, unlike the `while` loop, there is no need to initialize the loop variable as it will get its first value inside the loop.

Decision-making

Decision-making syntax in Java supports three structures available in C/C++ and other languages. They are the `if/else` statement, the `switch` statement, and the ternary operator.

A simple `if` statement does not require an `else` block:

```
if (age >= 65) {
    designation = "Senior";
}
```

You can create an either/or expression using `else`:

```
if (age >= 65) {
    designation = "Senior";
} else {
    designation = "Adult";
}
```

You can simplify this example by using the ternary operator, which uses a question mark and a colon:

```
String designation = (age >= 65) ? "Senior" : "Adult";
```

It begins with the logical test. While using parentheses in this situation is optional, I strongly recommend using them. After the question mark and on either side of the colon are the values that will be returned by the expression. You may also call a method if it returns a value of the appropriate type.

Should you need to define a test for ranges of value, you can use the `if/else/if` syntax:

```
if (age < 12) {
    designation = "child";
} else if (age < 18) {
    designation = "teenager";
} else if (age < 25) {
    designation = "young adult";
} else if (age < 65) {
    designation = "adult";
} else {
    designation = "senior";
}
```

Next up is the C-style switch. As of Java 17, the syntax of the C-style switch can be considered obsolete. The fact that the new versions of the switch are recent additions makes it important that you understand the C-style version. A switch is a logical structure for comparing the switch variable to the following:

- A literal integer
- An integer constant variable
- A literal string

Here is a switch to determine the postal rate, which depends on the zone the mail is being sent to:

```
double postage;
int zone = 3;
switch (zone) {
    case 1:
        postage = 2.25;
        break;
    case 2:
        postage = 4.50;
        break;
    case 3:
        postage = 7.75;
        break;
    default:
        postage = 10.00;
}
```

The lines that end in a colon are referred to as conditional labels. Should the zone variable's value match the literal, then the code that follows the matching case is performed. When such a match with a case is found, all subsequent cases become true, regardless of the case value. Therefore, there is a break statement at the end of every case. You can simulate a test against a limited range of values by purposely not using a break everywhere, as illustrated in the following code snippet:

```
String continent;
String country = "Japan";

switch (country) {
    case "UK":
    case "France":
    case "Germany":
```

```
            continent = "Europe";
            break;
        case "Canada":
        case "USA":
        case "Mexico":
            continent = "North America";
            break;
        default:
            continent = "Not found";
}
```

As of Java 14, two new versions of the switch were introduced. These are the new `switch` expression and new `switch` statement. This will also be the first time we see the new arrow operator. Here is the expression version of the switches:

```
postage = switch (zone) {
    case 1 -> 2.25;
    case 2 -> 4.50;
    case 3 -> 7.75;
    default -> 10.00;
};
```

The break is gone as any match will end the switch. To match one of multiple items, we can use the comma operator to create a list. The arrow operator (- >) points at the value that will be assigned to `continent`:

```
continent = switch (country) {
    case "UK", "France", "Germany" -> "Europe";
    case "Canada", "USA", "Mexico" -> "North America";
    default -> "Not found";
};
```

Unlike a `switch` expression, a `switch` statement does not return a value, but the matching case performs some action such as calling a method:

```
switch (continent) {
    case "Europe":
        showEuropeMap();
        break;
    case "North America":
```

```
            showNorthAmericaMap();
            break;
        default:
            showNotFound();
    }
```

Here is the new switch statement:

```
    switch (continent) {
        case "Europe" -> showEuropeMap();
        case "North America" -> showNorthAmericaMap();
        default -> showNotFound();
    }
```

There is another type of switch that, as of this writing, is only available as a preview feature in Java 19, and that is the pattern-matching switch. As a preview feature, it may change when it becomes formally part of the language or even dropped from the language. I see this as an exciting new type of switch—you can see it in action here:

```
    String designation;
    Object value = 4;
    designation = switch (value) {
        case Integer I when i < 12 ->
            "child";
        case Integer i when i < 18 ->
            "teenage";
        case Integer i when i < 25 ->
            "young adult";
        case Integer i when i < 65 ->
            "adult";
        default ->
            "senior";
    };
```

Pattern matching will only work with objects and not primitives unless they are in one of Java's primitive wrapper classes. When we assign the value 4 to the variable value of type Object, the compiler will auto-box the int primitive into an object of type Integer. Each case statement uses the class type in the case rather than a literal value. It also allows you to assign an identifier—in our case, i. Following the identifier is the new when keyword, after which you can write any valid Boolean expression. Only if the type matches and the logical expression after the when keyword is true is the case true. This should reduce the number of if/else if/ else if/ else if structures in your program. You will need Java 19 installed on your computer to experiment with this preview feature.

With how Java handles decisions out of the way, we can now look at how Java handles exceptions.

Handling exceptions

In Java, when things go wrong, they can be classified as errors or exceptions. An error is a problem that cannot be recovered from. An exception is an error that can be detected in your code such that you can possibly recover from it. For example, a recursion that never ends will result in a StackOverflowError-type error. Converting the Bob string to an integer will result in a NumberFormatException exception.

Here is a diagram of the primary exception classes:

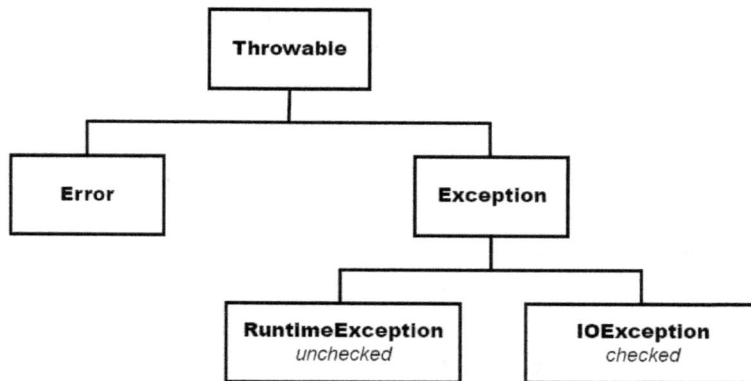

Figure 7.1 – The exception hierarchy

Exceptions are objects of classes named after the type of exception that has occurred. In the diagram, you can see that at the root of the hierarchy is the Throwable class. From Throwable, we have two subclasses: Error and Exception. The subclasses of Error are named after the errors that may occur during program execution. These are errors that generally cannot be recovered from and should lead to the program ending.

As it may be possible to recover from an exception as opposed to an error, these types of problems belong to the Exception branch. This branch is divided into two categories: **checked** and **unchecked**. A checked exception must come from a block of code called try/catch. Failing to use a try/catch block will generate a compiler error. You must resolve this; otherwise, you cannot compile your code.

Unchecked exceptions do not require a try/catch block. The compiler will happily compile code that might generate an unchecked exception without this code in a try/catch block. Should you decide not to handle an unchecked exception, your program will end.

Let's look at code that could have both types of exceptions:

```
public class FileException {
2
3       public void doCheckedException() {
4           List<String> fileContents = null;
5           Path path2 = Paths.get("c:/temp/textfile.tx"");
6           try {
7               fileContents = Files.readAllLines(path2);
8                   System.cut.printf("%s%", fileContents);
9           } catch (NoSuchFileException ex) {
10              ex.printStackTrace();
11          } catch(IOException ex) {
12              ex.printStackTrace();
13          }
14      }
15
16      public void doUncheckedException() {
17          int dividend = 8;
18          int divisor = ;
19          int result = dividend / divisor;
20          System.out.printf("%d%", result);
21      }
22
23      public void perform() {
24          checkedException();
25          uncheckedException();
26      }
27
28      public static void main(String[] args) {
29          new FileException().perform();
30      }
31  }
```

Let's review the important code lines.

Line 3 is the first method that contains code that could result in a checked exception being thrown.

Line 4 declares a `List` variable and sets it to `null`, which sets the `List` reference to zero. Local variables are not initialized, so they may already contain a value based on where in memory the reference is placed. If you do not properly allocate the `List` reference, usually done with `null`, there will be a compiler error in *line 11*. This will end the program.

Line 5 defines a path to a file. The `Paths.get()` method does not verify that the file exists, so no exception is thrown if the file does not exist.

Line 6 is the beginning of our `try` block where any code that may throw a checked exception is written. You may have lines of code in a `try` block that do not throw an exception.

In *line 7*, using `Files.readAllLines()`, each line in the file is added to the `List` variable. This is where an invalid file `Path` object can result in a checked exception named `IOException`.

Line 8 is the end of the `try` block and the beginning of the first `catch` block. A `catch` block takes as a parameter a reference to an `Exception` object that is created by the JVM when the exception is detected in the code inside the `try` block while the program executes. The `NoSuchFileException` exception is a subclass of `IOException`. Subclass exceptions must be handled before the superclass exception.

Line 9 is the body of a `catch` block, where you can write code to handle the error in such a way that the program does not need to end. All `Exception` objects have a method that displays the stack trace. You will not handle errors this way in production. When we discuss logging in the next chapter, we will see a best-practice approach.

In *line 10*, we have a second `catch` block. This is the `IOException` class. The code that reads the file can throw either a `NoSuchFileException` exception or an `IOException` exception. Some programmers may just catch `IOException`. As `NoSuchFileException` is a subclass of `IOException`, polymorphism allows you have both exceptions caught in one `catch` block that expects `IOException`. My preference is to use specific exception classes where possible.

Just as in *line 9*, here, in *line 11*, we are just printing the stack trace if this exception is caught here.

It is in *line 13* that a compiler error can occur if the `fileContents` variable is either not assigned `null` or assigned a reference from calling the `File.readAllLines` method.

During development, the use of the `printStackTrace` method in an `Exception` object can be useful. When we move to production code, we should never make this method call. In the next chapter, we will see how to use logging to preserve this information without it appearing in the console.

Line 16 is a method that will perform division by zero. This will generate an unchecked `ArithmeticException` exception. For this reason, you are not required to use a `try/catch` block. As the code is dividing by zero, an exception will be thrown, a stack trace will appear if this is a console application, and the program will end. A GUI program has no place to show a stack trace, so it will appear to just end suddenly.

The stack trace

When your program ends due to an exception or after catching an exception, you can display a stack trace. The stack trace will appear in the console window. It is a list of every line of code that led to the exception before being caught or after the program ends. Here is the stack trace from the doCheckedException method when the filename in the Path object cannot be found:

```
Command Prompt          x   + ~                                                    –   □   ×

C:\PacktJavaCode>java FileException.java
java.nio.file.NoSuchFileException: c:\temp\textile.txt
        at java.base/sun.nio.fs.WindowsException.translateToIOException(WindowsException.java:85)
        at java.base/sun.nio.fs.WindowsException.rethrowAsIOException(WindowsException.java:103)
        at java.base/sun.nio.fs.WindowsException.rethrowAsIOException(WindowsException.java:108)
        at java.base/sun.nio.fs.WindowsFileSystemProvider.newByteChannel(WindowsFileSystemProvider.java:236)
        at java.base/java.nio.file.Files.newByteChannel(Files.java:380)
        at java.base/java.nio.file.Files.newByteChannel(Files.java:432)
        at java.base/java.nio.file.spi.FileSystemProvider.newInputStream(FileSystemProvider.java:422)
        at java.base/java.nio.file.Files.newInputStream(Files.java:160)
        at java.base/java.nio.file.Files.newBufferedReader(Files.java:2921)
        at java.base/java.nio.file.Files.readAllLines(Files.java:3411)
        at java.base/java.nio.file.Files.readAllLines(Files.java:3452)
        at FileException.doCheckedException(FileException.java:17)
        at FileException.perform(FileException.java:63)
        at FileException.main(FileException.java:69)
        at java.base/jdk.internal.reflect.DirectMethodHandleAccessor.invoke(DirectMethodHandleAccessor.java:104)
        at java.base/java.lang.reflect.Method.invoke(Method.java:578)
        at jdk.compiler/com.sun.tools.javac.launcher.Main.execute(Main.java:434)
        at jdk.compiler/com.sun.tools.javac.launcher.Main.run(Main.java:205)
        at jdk.compiler/com.sun.tools.javac.launcher.Main.main(Main.java:132)

C:\PacktJavaCode>
```

Figure 7.2 – The stack trace explicitly displayed when an exception occurs

As you can see, the exception has traveled through several methods, many of which occurred in a Java library and not your code. To use this information to locate the possibly offending source code, go through the list and locate the first entry that comes from your code, starting from the beginning of the trace:

```
at com.kenfogel.FileException.doCheckedException(FileException.
java:17)
```

This line tells us that the exception happened in the doCheckedException method on *line 17*.

Ending the program

In some situations, you may wish to end a program after catching its exception. You can end most programs with System.exit(n), where n is a number you assign to this error:

```
        } catch(IOException ex) {
            ex.printStackTrace();
            System.exit(12);
        }
```

The number—in this example, 12—maps to a known error condition that must end the program. Here, after the stack trace is displayed, the program ends.

The throw and throws statements

If an exception is thrown in a method, Java looks for a catch block. If there is no catch block in the method that threw the exception, then Java looks into the method that called the offending method. This continues until it gets to the main method, and at that point, the program ends. There are situations where you will want to catch an exception where it happens, but then you want to re-throw it to whatever method that came before it that has a catch block. In this catch block, we are displaying the stack trace to the console and then re-throwing the exception:

```
} catch(IOException ex) {
    ex.printStackTrace();
    throw ex;
}
```

To be able to re-throw, we must add to the method a throws clause:

```
public void doCheckedException() throws IOException {
```

When you use throws, whichever method calls doCheckedException must do this in a try/catch block, as shown:

```
try {
    checkedException();
} catch(IOException ex) {
    ex.printStackTrace();
}
```

We can also use the throws clause to define that a method has a checked exception, but it will not be handled in the method. This means we can just call checkedException() without a try/catch block, as the method shows that it will be thrown to whichever try/catch block in another method called this method.

The finally block

There can be a third block for handling exceptions, called the finally block. In this block, you can write any code that you wish to execute if an exception is thrown or not. In this example, a message is displayed regardless of whether an exception is thrown or not:

```
public void doFinallyExample(int dividend, int divisor) {
```

```
        int result = 0;
        try {
            result = dividend / divisor;
        } catch (ArithmeticException ex) {
            ex.printStackTrace();
        } finally {
            System.out.printf(
                "Finally block is always executed%n");
        }
    }
}
```

If the divisor is valid—not zero—then the code in the `finally` block is executed. If the divisor is invalid—is zero—the code in the `catch` block is executed followed by the code in the `finally` block.

> **Note**
>
> Do not confuse `finally` with the `finalize` method. The `finally` block is useful. The `finalize` method is not useful and should not be used.

Creating your own exception classes

The name of an exception class is the description of the problem that led to the exception being thrown. You can create your own exceptions and then throw your custom exception when you detect a serious problem in your code. The first step is to create an exception class, like so:

```
public class NegativeNumberException extends Exception{}
```

This is a checked exception class. If you do not want it to be checked, then extend `RuntimeException`. You can add additional methods or override methods in `Exception`, but this is not necessary. You create custom exceptions to define exceptions unique to your program that are not sufficiently described in the existing family of exception classes.

Now, we need some code that will throw this exception:

```
    public void doCustomException(int value)
                    throws NegativeNumberException {
        if (value < 0) {
            throw new NegativeNumberException();
        }
        System.out.printf("value = %d%n", value);
    }
```

Now, we need code that will call this method. As the method we are calling has a `throws` clause, we must treat it as a checked exception, and we must use a `try`/`catch` block:

```java
public void makeCustomException() {
    try {
        doCustomException(-1);
    } catch (NegativeNumberException ex) {
        ex.printStackTrace();
    }
}
```

Here is the stack trace that occurred when this code executed:

```
com.kenfogel.NegativeNumberException
   at com.kenfogel.FileException.
doCustomException(FileException.java:39)
   at com.kenfogel.FileException.
makeCustomException(FileException.java:46)
    at com.kenfogel.FileException.perform(FileException.java:69)
    at com.kenfogel.FileException.main(FileException.java:73)
```

You can see that the exception class we created is the exception reported in the stack trace.

There is one last issue to point out in regard to exceptions. Many languages such as C# and JavaScript do not have checked exceptions. The decision to catch these exceptions is solely at the discretion of the developer.

Throwing an exception is a slow process in the JVM. It is not something you might notice, but if it happens often enough, it will result in slower execution of the program. For this reason, never use exception handling as part of the program logic. Exceptions are serious issues that, in most cases, imply an error or bug that can affect the outcome of the program. If you can detect an error in your code, typically by testing a value with an `if` statement, you should handle it with the code you write and not by expecting or throwing an exception.

Summary

In this chapter, we learned about how Java code is organized into blocks as defined by an opening and closing brace. The blocks can be an entire class, each method in the class, and a body of iteration and decision statements. From there, we learned how to classify lines of code as statements or expressions.

Operators were the next topic. We reviewed the math and logic operators and how they are combined. The `cast` operator for converting from one type to another was also shown.

Next up were the two most common coding structures: iterations and decisions. The classic `for` loop, a loop where the number of iterations is known before the loop begins, was presented. The second style of loops was `while` and `do/while` loops. These loops do not know how many iterations there will be. This is determined in the repeating block of code.

Decision-making was next up. We looked at the `if` and `if/else` statements. These are effectively the same as found in any language that traces its lineage to the C language. The second decision structure we covered was the `switch` statement. As with `if`, it is virtually unchanged from its C roots. The good news is that this style of switch has been enhanced with three new versions.

The last topic we looked at was exceptions. We looked at what exception classes and objects are and which category, checked or unchecked, they fall into. How we can handle exceptions rather than just let the program end was presented as well. Creating our own named exceptions and how we can use them was the last topic we covered.

At this point, you should feel comfortable reading Java code. In our next chapter, we will look at additional features of the language and how they can be used to write cleaner code.

Further reading

- *Exceptions in Java*: https://medium.com/interviewnoodle/exception-in-java-89a0b41e0c45

8

Arrays, Collections, Generics, Functions, and Streams

Up until now, we have used variables to represent a single instance of a primitive data type and a reference data type. More frequently encountered in the real world, though, is the need to work with and process multiple data elements. In this chapter, we will look at the various options available for managing multiple elements. In examining the options in Java for this purpose, we will see how we can enhance type safety.

To process multiple elements more efficiently, we will examine streams—a replacement for traditional loops when coupled with functions.

We will cover the following topics:

- Understanding the array data structure
- Understanding the Collections Framework
- Using sequential implementations and interfaces
- Understanding Generics in the Collections Framework
- Using sequential implementations and interfaces with Generics
- Understanding Collections Framework map structures
- Understanding functions in Java
- Using streams in collections

Upon completion of this chapter, you will be able to work with multiple instances of data as an array or collection and apply algorithms available in the Stream library.

Technical requirements

Here are the tools required to run the examples in this chapter:

- Java 17 installed
- Text editor
- Maven 3.8.6 or a newer version installed

The sample code for this chapter is available at `https://github.com/PacktPublishing/ Transitioning-to-Java/tree/chapter08`.

Understanding the array data structure

As with most languages, Java has a built-in array data structure and does not require any imports or external libraries. As such, the array behaves as most arrays in other languages. The only difference is that to instantiate an array, you need the new keyword. Here are the two ways to declare an array of 10 elements of type `int`:

```
int[] quantities01 = new int[10];
int quantities02[] = new int[10];
```

The difference is where the empty square brackets are placed on the left-hand side. Placing them after the type is considered the Java way. Placing it after the identifier is thought of as the C-language way. Either syntax is fine.

In most programming languages, numbers can be either ordinal or cardinal. The length of the array as declared when we instantiate it is a cardinal—or count—number. In the examples so far, the length has been 10. An ordinal number represents the position in a structure such as in an array. Ordinals begin with zero in most programming languages. When we declare an array of cardinal length 10, the ordinal positions range from 0 to 9 and not 1 to 10.

Arrays are of a fixed length; they cannot be expanded or contracted. Each position in an array is ready for use. You can assign a value to the last position before assigning one to the first position.

You can defer instantiating an array, as follows:

```
int[] quantities01;
...
quantities01 = new int[10];
```

Arrays store the values of an array in a contiguous block of memory. This tells us that the array will be 10 elements times 4 bytes per `int` to consume 40 bytes plus the necessary overhead for an object in Java. The length of the array is part of this overhead.

An array of objects consists of an array of references. For example, you could create an array of four strings, like so:

```
String[] stuff = new String[4];
```

In this case, the array will consume 4 bytes for the reference to each String object as well as the usual array object overhead. The strings themselves are stored in memory as the **Java Virtual Machine (JVM)** decides and are not necessarily contiguous. The value of each reference will be null until you assign a valid reference to the array:

```
String myThing = "Moose";
stuff[0] = myThing;
```

When using references in arrays, what we are storing in the data structure is the reference and not the object. Only primitives can be stored directly in an array.

From here, we read and write to the array using the subscript. To get the length of the array, we use the final constant variable, length:

```
System.out.printf("Length: %d%n", stuff.length);
```

To visit every element in an array with a for loop, you would use the following:

```
stuff[0] = "Java";
stuff[1] = "Python";
stuff[2] = "JavaScript";
stuff[3] = "C#";

for (int i = 0; i < stuff.length; ++i) {
    System.out.printf("Stuff %d = %s%n", i, stuff[i]);
}
```

Java also has an enhanced for loop for visiting every element in an array. The subscript value is no longer available:

```
for(String s : stuff) {
    System.out.printf("Stuff %s%n", s);
}
```

One last thing to point out: Java has a library for performing a range of operations on an array, called the Arrays library. This class contains static methods for sorting and searching as well as creating a list, one of the collections, from an array. We will see an example of turning an array into a list in the later section, *Using streams in collections*.

You should already be comfortable with working with an array. You can read and write to any valid subscripted element. If you use an invalid subscript that is out of range, Java will throw an `ArrayIndexOutOfBoundsException`. Now, let's look at the Collections Framework.

Understanding the Collections Framework

Once an array is instantiated, it cannot have its length increased or decreased. This means that you must know the exact number of elements you will need before you instantiate the array. You can use a variable to declare the array but once created it cannot be resized. Have a look at the following example:

```
int numberOfCats = 6;
int[] cats = new int[numberOfCats];
```

This is where collections come in. These are dynamic data structures that can increase in size as elements are added. You can also remove elements, although reducing the size is not always available, and if it can be reduced, then you must call an appropriate method.

The Collections Framework is divided into implementations and interfaces. An implementation may support more than one interface. While an implementation can have a large selection of methods, the use of an interface allows you to restrict what you can do with a collection.

The Collections Framework classes fall into two categories. There are sequential collections that preserve the order in which elements are added. Then, there are map collections where elements are stored in pairs of data. The first is typically a field of an object called the key, while the second is a reference to the object itself called the value. These collections organize themselves based on the key.

The default data type that all members of these classes manage is `Object`. This means that you can store any object in a collection as all classes extend `Object`, and polymorphism allows you to use a subclass wherever a superclass is called for. The problem with this approach is that you can conceivably have a collection of apples and oranges. Until the introduction of Generics to the language, it was the responsibility of the developer to avoid mixing types.

Let's look more closely at sequential structures.

Using sequential implementations and interfaces

Let's begin with the implementation. These are classes that manage the data in many ways. They are `ArrayList`, `LinkedList`, and `ArrayDeque`.

ArrayList

This is a dynamic array-like structure. As a class, you must use methods rather than subscripts to access specific elements. You add elements at the end of the list. Once you add an element, you can read from it, write to it, search for a specific value, and remove elements from a specific position or that match a specific value.

You can instantiate an `ArrayList` class with or without an initial capacity. If you do not specify a capacity, then it will default to a capacity of 10. If you know in advance how many elements you will need, then include that value when you instantiate the `ArrayList` class. The auto-resizing of an `ArrayList` class entails overhead that you can avoid if you know the precise size. In either case, you cannot access elements until you first add an element. As you add elements, the size increases. You can access any element that you add but you cannot access any positions between the last element added and the unused capacity that follows it.

LinkedList

This structure stores data in node objects with each node knowing what comes before and after it. On the surface, it would seem to be quite efficient as you create nodes as needed. The major drawback to a linked list is that it does not support random access. In `ArrayList`, you can access any element such as an array by using the integer that represents its position using a method rather than square brackets. This access is direct. In a `LinkedList` class, the only elements you can access directly are the first and last elements. To access any other element, you must start at the beginning or end and then follow the forward or backward references to the subsequent nodes. This makes access to elements far slower than in `ArrayList`.

I instruct students about linked lists because they make for nice blackboard diagrams. The `Map` structures that we will look at shortly are based on variants of the linked list. Let me end with a tweet from Joshua Bloch concerning the Java `LinkedList` class he wrote:

Figure 8.1 – Famous LinkedList tweet

ArrayDeque

The `ArrayDeque` class is like `ArrayList` in that it is a dynamic structure that stores elements in an array-like structure. Unlike `ArrayList`, it does not support direct access. Instead, it is optimized for inserting or removing elements at the beginning (**FIFO**) or at the end of the structure (**LIFO**). This leads to the data structures defined by the `Deque`, `Queue`, and `Stack` interfaces. Prior to the introduction of the `ArrayDeque` class in Java 1.6, you used a `LinkedList` class as the implementation for these interfaces. The `ArrayDeque` class outperforms the `LinkedList` class.

This is not a complete list. For example, there is a `Stack` class, but using the `ArrayDeque` class with a `Deque` interface will outperform the `Stack` class. A second issue relates to thread safety. These three implementations are not thread-safe. There are specialized versions of the implementations in the framework, specifically for when threads must share access to a data structure.

You can implement any of these classes, but it is considered a poor choice. Each of these implementations has numerous methods to support the use of the structure in numerous ways. When you use a Java collections class, you want to use the smallest interface for what you are trying to accomplish rather than allowing access to every method in the implementation. Let's look at these interfaces.

The Collection interface

Here is a diagram of the most common interfaces:

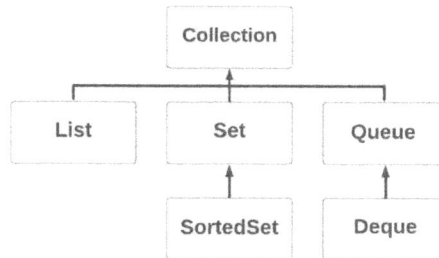

Figure 8.2 – The Collection interfaces

Each of the boxes represents an interface that a collection implementation may or may not support. Collection is the super interface. Any class that implements the interfaces below it must also implement Collection.

The most common interface is List. This is the closest to an array. Set and SortedSet are interfaces that ensure that an element cannot appear more than once. Queue is a FIFO structure. You can only add elements to the end of the structure, and you can only remove them from the front of the structure. Deque is a structure that supports LIFO. What is unique about Deque is that you add or remove from either end. Neither Queue nor Deque permit access by subscript.

How to declare a collection

As we have already discussed in *Chapter 6, Methods, Interfaces, Records, and Their Relationships*, you use an interface class to define which methods the class must implement. The most widely used interface for sequential collections is the List. We can now declare a data structure that can only use the methods shown in the List interface and no others:

```
List moreStuff = new ArrayList();
```

Before we look at more interfaces, it is time to look at the concept of Generics and how they relate to the collections interface. We need to look at this now because while the previous line of code is executable, rarely will a Java developer write it this way.

Understanding Generics in the Collections Framework

As pointed out, the default classes in the Collections framework were designed to manage only references to objects of type `Object`. Polymorphism then permitted the use of any subclass in an implementation of any of these classes. The problem with this approach is that it is not type-safe. Look at this code fragment:

```
int numberOfApples = 9;
String orange = "Valencia";


List stuff = new ArrayList();
stuff.add(numberOfApples);
stuff.add(orange);


System.out.printf("Stuff: %s%n", stuff);
```

This code begins by declaring two variables. The first is an `int` type with the `numberOfApples` identifier. Collections cannot contain primitive data types, so an object of type `Integer` is required if the primitive is an `int` type. Java will perform this conversion from primitive to object for you. The second line creates a `String` object.

Next is the instantiation of an object of type `ArrayList` but whose interface is restricted to just what the `List` interface class permits. Now, we can add the `Integer` and the `String` objects to the collection. The last line displays the contents of the `List` as its `toString()` method creates a `String` object of all members. This leads to the expression that you should not mix apples and oranges. Collections must be of a single type. While the default syntax for Collections does not restrict what can be added, the use of Generic notation will.

Let's look at a new variation of the previous code:

```
int numberOfApples = 9;
String orange1 = "Valencia";
String orange2 = "Navel";


List<String> stuff = new ArrayList<>();
stuff.add(orange1);
stuff.add(orange2);
stuff.add(numberOfApples);


System.out.printf("Stuff: %s%n", stuff);
```

In this example, we kept our `int` type and then created two strings. The declaration of the `List` now includes angle brackets. Within the brackets is the class type you want to restrict the `List` to contain. In this example, the class is `String`. While we must show the class type on the left-hand side, we can just have empty angle brackets on the right side as there is never a situation where these two class types could be different.

The next lines add the objects to the `List`. The first two will work, but the third, where we are trying to add an object of type `Integer`, will generate an exception:

```
java.lang.RuntimeException: Uncompilable code - incompatible
types: java.lang.Integer cannot be converted to java.lang.
String
```

Java will no longer allow you to mix apples and oranges. This test for ensuring all objects added to a collection are the same type only occurs at compile time. This means it is possible to add a different object type if this operation only occurs at runtime. This can occur when multiple processes are running in the JVM, and one process calls a method in another process.

We will now return to the collections and only use Generic syntax from here on in.

Using sequential implementations and interfaces with Generics

As we have just seen, the best practice for creating a `List` will be this:

```
List<String> moreStuff = new ArrayList<>();
```

Any valid class type can be used. Once we have elements in our collection, we can access them with the `get` method and the subscript:

```
String orange1 = "Valencia";
String orange2 = "Navel";

List<String> stuff = new ArrayList<>();
stuff.add(orange1);
stuff.add(orange2);

System.out.printf("Stuff: %s%n", stuff.get(0));
System.out.printf("Stuff: %s%n", stuff.get(1));
```

In the last two lines, we are referring to specific positions in the list. To change the object stored in a specific position, we use the set method:

```
stuff.set(0, "Blood Orange");
```

The interfaces and the implementation classes all support methods that allow you to determine if a specific object is contained in the collections. For this reason, you must override the equals method inherited from Object. Certain collection methods require the hash value, so your classes must have a hashCode method.

You can also sort collections. To do this, the class type of the objects you are storing must implement the Comparable interface. This interface requires you to write just one method named compareTo that returns a negative number, zero, or a positive number. Here is a fragment of a class that implements compareTo:

```
public class ComparableClass implements
                        Comparable<ComparableClass>{
    private final int value;
    public ComparableClass(int initialValue) {
        value = initialValue;
    }
    @Override
    public int compareTo(ComparableClass o) {
        return value - o.value;
    }
}
```

This class has only one field that is initialized by the constructor. It implements the Comparable interface using Generic notation to indicate that we can only compare this object to an object of the same class.

The compareTo method, required because we are implementing the Comparable interface, must return:

- A positive integer if the value of the current object we are comparing is greater than the value of the object that is being compared to
- The value 0 if the value of the current object we are comparing is equal to the value of the object that is being compared to
- A negative integer if the value of the current object we are comparing is less than the value of the object that is being compared to

You may wonder how we can access the value of ComparableClass that is passed to compareTo using dot notation when the field is private. This is possible because Java permits objects of the same class to access private members of another instance of this same class. Here is the class that tests this:

```java
public class ComparableTest {
    private final ComparableClass comparable01;
    private final ComparableClass comparable02;

    public ComparableTest(int value1, int value2) {
        comparable01 = new ComparableClass(value1);
        comparable02 = new ComparableClass(value2);
    }
    public void perform() {
        System.out.printf("comparable01 to comparable02 %d%n",
                comparable01.compareTo(comparable02));
        System.out.printf("comparable02 to comparable01 %d%n",
                comparable02.compareTo(comparable01));
    }
    public static void main(String[] args) {
        ComparableTest examples = new ComparableTest(12, 2);
        examples.perform();
    }
}
```

In the perform method, we are displaying the result of calling the compareTo method. Let's now create a List of objects and then sort the list. There is a small modification to ComparableClass. A method to return the value stored in the class has been added:

```java
public int getValue() {
    return value;
}
```

Now, we have a class that creates 10 objects of ComparableClass, places them in a List, and sorts the List:

```java
public class ComparableSorting {
    private final List<ComparableClass> comparableClasses;
```

Here is the constructor that instantiates an object of type `ArrayList` that will be restricted to use only methods from the `List` interface

```
public ComparableSorting() {
    comparableClasses = new ArrayList<>();
}
```

Note that the angle brackets after `ArrayList` are empty. When we declared `comparableClasses`, we declared the `List` as containing `ComparableClass` objects. There is no need to repeat this.

This next method creates 10 instances of `ComparableClass`, initializing them with a random integer as they are added to the `List`. Each value is also displayed on the console so that we can see the original values as they are assigned:

```
private void fillList() {
    Random rand = new Random();
    int upperBound = 25;
    System.out.printf("Unsorted:%n");
    for(int i = 0; i < 10; ++i) {
        comparableClasses.add(
            new ComparableClass(rand.nextInt(upperBound)));
    }
    System.out.printf("%n");
}
```

This method displays the values in each object in the `List`:

```
private void displayList() {
    for(int i = 0; i < 10; ++i) {
        System.out.printf(
            "%s ", comparableClasses.get(i).getValue());
    }
    System.out.printf("%n");
}
```

Now, let's fill the list, display it, sort it, and display it again:

```
public void perform() {
    fillList();
    displayList();
```

The `Collections` class contains a family of static methods that can be applied to an object that implements the `Collection` interface. One is the `Collections.sort` method. It alters the input rather than returning a new value:

```
        Collections.sort(comparableClasses);
        displayList();
    }
    public static void main(String[] args) {
        ComparableSorting examples = new ComparableSorting();
        examples.perform();
    }
}
```

What we have seen so far is how sequential collections can be used. The need for special interfaces such as `Comparable` was also highlighted. Let's now look at ordered collections.

Understanding Collections Framework map structures

The second family of collections is the map family. A map is a data structure in which you add elements to a map with a pair of values. The first value is the key. This is a reference to an object that, depending on the type of map, either implements the `Comparable` interface—as we saw in the previous section—or overrides the `hashCode` and `equals` methods. If the key is a primitive, then we declare it as its wrapper class, and Java will manage the necessary conversion to and from the primitive. The second is the value—a reference to the object you are storing in the map. This class does not need to implement the `Comparable` interface.

There are three map implementations in Java, which we will now cover.

HashMap

Of all the data types available in Java and most other languages, the fastest performing is the integer. The size of an integer is the same as the word size of the CPU. The JVM is a 32-bit or 4-word machine. Even 64-bit Java is just modeling a 32-bit machine. This is where the hash code comes in.

As with all map structures, entries are comprised of two components. The first is the key and the second is the value. What makes `HashMap` special is that the hash code of the key value determines where it will store the pair. The underlying structure is an array, and each position in the array is a bucket. Using arithmetic operations such as modulus, the subscript in the array can be determined from the hash code of the key.

A hash code is not unique. This means that two or more keys may generate the same hash code. In this case, they will also want to use the same subscript, and the bucket becomes a single linked list of buckets. If the number of keys becomes greater than eight, then the linked list is converted into a balance binary tree. When searching for a key in a list, the `equals` method is used to test each bucket to find the value.

Use a hash map when you must collect data that you must be able to retrieve from the structure rapidly. There is no defined order. The first item you put in a hash map could be the seventh element in the array of buckets. This also means that the order of elements put into the structure cannot be determined.

To find a value in a `HashMap` object given a key, you can use the `get` method. This method takes the key as a parameter and returns the value if found or it returns `null` if not found. Let's look at an example.

First, we create a `HashMap` object using the `Map` interface:

```
Map<Integer, Integer> hashMap = new HashMap<>();
```

Now, we can put data into the `HashMap` object using the `put` method that takes two parameters. These are the key and the value:

```
hashMap.put(5, 6);
hashMap.put(5, 4);
hashMap.put(4, 8);
hashMap.put(3, 10);
hashMap.put(2, 6);
```

These next two lines will retrieve the value associated with the key if the key exists. Otherwise, `null` is returned:

```
System.out.printf("%s%n",hashMap.get(4));
System.out.printf("%s%n",hashMap.get(1));
```

There is no entry in the `HashMap` object that uses the integer 1 as a key, so it will print out `null`.

To iterate or work with every element over the entire `HashMap` object, we need to first create a `Set` object from all the entries in the hash map:

```
Set s = hashMap.entrySet();
```

From the `Set` object, we create an `Iterator` object. An iterator allows us to access every element in the set in the order determined by the keys:

```
Iterator it = s.iterator();
```

The `Iterator` object's `hasNext` method returns `true` if there is another element in the `Set` object; otherwise, it returns `false`:

```
while (it.hasNext()) {
```

The `Iterator` object's `next` method returns the key/value pair:

```
        System.out.printf("%s%n",it.next());
    }
```

The output of this code will be this:

```
8
null
2=6
3=10
4=8
5=4
6=6
```

Notice that the order of keys is not the same as the order they were put into the hash map.

LinkedHashMap

This structure is a variant of `HashMap`. Internally it operates just like `HashMap` but also includes a second data structure. This is a linked list that preserves the order in which data is put into `LinkedHashMap`. If the order of entry is not significant, use a `HashMap` structure.

If we used `LinkedHashMap` in the previous example code, the only change we would make would be to use `LinkedHashMap` instead of `HashMap`:

```
        Map<Integer, Integer> linkedHashMap = new
            LinkedHashMap<>();

        linkedHashMap.put(6, 6);
        linkedHashMap.put(5, 4);
        linkedHashMap.put(4, 8);
        linkedHashMap.put(3, 10);
        linkedHashMap.put(2, 6);

        Set s = linkedHashMap.entrySet();
```

```
Iterator it = s.iterator();
System.out.printf("key=Value%n");
while (it.hasNext()) {
    System.out.printf("%s%n",it.next());
}
```

The output of this version will be this:

```
6=6
5=4
4=8
3=10
2=6
```

This is the same order that the key/value pairs were put into the map.

TreeMap

Unlike HashMap and LinkedHashMap, the underlying structure of TreeMap is a red-black binary tree. The key value is used as is and must implement the Comparable interface. You do not need the hashCode and equals methods, but it is good practice to include them. Here is the same code using TreeMap:

```
Map<Integer, Integer> treeMap = new TreeMap<>();
```

Here, the keys are not in any order. As integers, they do have a natural order that will determine where in the binary tree the key/value pairs are placed:

```
treeMap.put(6, 6);
treeMap.put(4, 4);
treeMap.put(3, 8);
treeMap.put(2, 10);
treeMap.put(5, 6);
```

When we use the iterator to display all the key/value pairs, they will be in the order based on the key:

```
Set s = treeMap.entrySet();
Iterator it = s.iterator();
while (it.hasNext()) {
    System.out.printf("%s%n",it.next());
}
```

The output will be this:

```
2=6
3=10
4=8
5=4
6=6
```

While an array is frequently the go-to structure when you need multiple elements, consider that its interface is quite limited. Collections have a rich set of methods that expand what you can do in your code. Before we move on to functions, keep in mind the following note.

> **Important note**
> The Collections shown in this chapter are not thread-safe. There are variants of each collection that are thread-safe.

Understanding functions in Java

In Java, we call units of code in a class a method. In C and C++, we call them functions. In JavaScript, we even use the keyword function. What sets Java apart from these other languages is that functions represent a different coding model than classes and their methods. There are functional rather than **object-oriented (OO)** languages, of which Haskell is one example. We are briefly examining functions because our next topic, streams, is based on the function rather than the class model.

Let's look at some code that attached an event handler to a button in JavaFX. We will be looking at JavaFX in *Chapter 13, Desktop Graphical User Interface Coding with Swing and JavaFX*. Let's begin by looking at what a functional EventHandler interface is:

```
@FunctionalInterface
public interface EventHandler<T extends Event> extends
    EventListener {
    void handle(T event);
}
```

This is the interface class for EventHandler that is part of JavaFX. The @FunctionalInterface annotation is optional but adds clarity to the purpose behind this interface. Functional interfaces can only have one abstract method. There is no implementation of this interface in JavaFX. You must supply the code for the handle method:

```
btn.setOnAction(new EventHandler<ActionEvent>() {
    @Override
```

```
        public void handle(ActionEvent event) {
            actionTarget.setText(userBean.toString());
        }
    });
```

This code registers an event handler for when a button is pressed. The handler must be an object of type EventHandler and must have a method named handle. The handle method is calling upon the toString method from userBean to return a string that will be assigned to a Text field named actionTarget.

The first fragment demonstrates an anonymous inner class. It is anonymous because the reference is never assigned to an identifier. It can only be used in the method call. We do this in situations where the action to be performed is unique for just this button press. It cannot be reused elsewhere:

```
    btn.setOnAction((ActionEvent event) -> {
        actionTarget.setText(userBean.toString());
    });
```

This second fragment uses lambda notation. There is only one method, handle, in EventHandler. Therefore, we do not require additional decorations. (ActionEvent event) is the parameter that the handle method must receive. Our lambda is providing the code for the handle method that will execute when the button is pressed. While there is no limit to the number of lines of code that can be in a lambda, the best practice is no more than three, while just one line is preferred. But what if multiple lines need to be executed? This leads us to the third syntax for using a function:

```
    btn.setOnAction(this::signInButtonHandler);
```

Functions in Java can be passed as a parameter to a method and can also be returned by a method. In this fragment, we are stating that the signInButtonHandler method will be invoked as if it were the handle method:

```
private void signInButtonHandler(ActionEvent e) {
    actionTarget.setText(userBean.toString());
}
```

Here, this method is in the same file. Therefore, we are referring to it with this in setOnAction. It must have the same return value and parameters as the handle method.

Functional programming helps to simplify our code. In any situation where we need a method whose action must be defined where it will be used, then using functions is the best choice.

Using streams in collections

Processing all data elements in a collection is a common action. Maybe you want to extract a subset of the collection based on a specific requirement. You might want to increase or decrease values or change the case of strings. This is where streams come in. All classes that implement the `Collection` interface have a stream method from which we can chain numerous stream methods. You cannot use streams directly on maps, but if you convert a map into a set, then you can use streams.

One important characteristic of stream methods is that they operate as pure functions. A pure function does not change the state of any fields in the class or any of the parameters passed to it. A stream method always returns a new stream. The original stream has not changed. Let's see how this works:

```
public record Employee(String employeeId, String firstName,
        String lastName, String department, double salary) { }
```

Here, we have a record that contains fields that represent information on an employee. Now, let's create a `List` of six employees. This information should come from a database, but for our purposes, we will create the `List` in the constructor of the `StreamsExample.java` file:

```
public class StreamsExample {
    private List<Employee> staffList;
    public StreamsExample() {
        staffList = Arrays.asList(new Employee("A9", "Benson",
                        "Bill", "Sales", 56000),
                new Employee("A1", "Clarkson",
                        "Bill", "Sales", 56000),
                new Employee("A2", "Blunt",
                        "Wesley", "HR", 56000),
                new Employee("A3", "Smith",
                        "Joan", "Software", 56000),
                new Employee("A4", "Smith",
                        "John", "Accounting", 56000),
                new Employee("A5", "Lance",
                        "Gilbert", "Sales", 56000));
    }
```

This code demonstrates that an array, not a collection, can be created by listing the elements separated by commas when declared in a block. There are six `Employee` objects created, and by using the `Arrays` method `asList`, they are converted into a `List`-compatible structure. This is necessary as streams do not work on arrays. With our list complete, we can now apply the methods available to streams. Many of the stream methods behave in a similar manner to SQL actions.

With the list in place, let's use some of the stream methods:

```
List<Employee> filteredList = staffList.stream().
    filter( s -> s.firstName().startsWith("J")).
    collect(Collectors.toList());
```

After we call upon `stream()`, we can apply the stream methods. The first one is `filter`. It requires a function that can return `true` or `false` depending on the code in the function. The function is expressed as a lambda that will receive a record object of type `Employee`. The code in the lambda retrieves the `firstName` string from the record and applies the `startsWith` string method to find names that begin with `"J"`.

The result from `filter` is a new stream with only the objects that match the criteria. A stream must be converted back into a collection, and that is the role of the `collect` method. It takes as its argument the `Collectors.toList()` function from the `Collectors` class that will return a `List` object.

One last example, and that is to print out the employee objects in order sorted by the `lastName` field:

```
staffList.stream().sorted((e1, e2) ->
    e1.lastName().compareTo(e2.lastName())).
    forEach(System.out::println);
```

This line of code uses the `sorted` stream function. It requires a function that determines the order of two objects based on a field. The field chosen is a `String` object, so it already has a `compareTo` method; otherwise, you will need to write a `compareTo` method. The lambda expression takes two parameters that are defined as `String` objects from the `Record`. This will produce a stream in sorted order that is then used by the `forEach` function. Rather than return a new `List` object, the `forEach` function receives a stream and passes each member of it to the `println` method of `System.out`.

Modern Java programs rely heavily on streams. The alternative is to use iterators or `for` loops to access all elements. If you need to process all elements in a collection, look at streams before anything else.

Summary

While we appear to have covered a lot of topics in this chapter, you should recognize that they are all related. Whether you are retrieving records from a database or receiving user input from the keyboard, as soon as there is more than one item, you will need what has been presented here. We began with a basic array that is like a structure in other languages. From the array, we moved on to the Collections Framework. These are dynamic structures that can grow as needed. From the sequential to the map collections, Java provides us with a rich set of choices.

We looked at Generics next. Unlike an array that is declared as a specific type, raw collections can store any object without regard to what has already been stored. Using generic notation, we can tie a collection to a specific data type.

Starting with Java 8, functions became part of the Java language. While ordinary methods in a class can be used as a function, the use of lambdas allows us to define specific actions for a particular problem. The Stream library, available in sequential collections, simplifies processing the elements of a collection.

Next up, we will examine how we can document our code and record information on the operation of our program in logs.

Further reading

- *Guide to the java.util.Arrays class*: `https://www.baeldung.com/java-util-arrays`
- *Collections in Java*: `https://www.scaler.com/topics/java/collections-in-java/`
- *Functional Programming in Java*: `https://www.scaler.com/topics/java/functional-programming-in-java/`
- *Java Stream*: `https://zetcode.com/java/stream/`

Using Threads in Java

One of the first software development contracts I had was to develop software for an invisible fence security system at thoroughbred horse farms in Kentucky, USA. The computer we used was an Apple II plus. There was no such thing as threads in the 6502 CPU or the OS, ProDOS. What we did was write all the code in assembly language in small units that were measured by the number of cycles each unit took. Once we finished our allotted cycles, we would save our state in a defined region of memory and turn over control to the next unit. It worked quite well, and if a horse wandered off, alarms would be sounded. The monitoring of the fence, just a buried cable that could detect a horse walking over it, continued even while the alarm sounded. This was my introduction to threaded programming. This was in 1982.

I did not work with threads again until 1999 when I moved from C++ to Java. One of the features that made Java stand out and why I abandoned C++ was Java's standardized support for threading in the language. This, along with support for GUI applications with Swing, made it clear to me that Java was the language that students in my program needed to learn. In 2002, Dawson College's Computer Science Technology program, of which I was the chairperson, abandoned COBOL as the primary teaching language in favor of Java.

Today, many languages, including C++, have native support for multithreaded programming. In this chapter, we will examine how you can write threaded code in Java that depends on the Java virtual machine working with the computer's OS. There are issues to deal with when threads share a resource. Using synchronization to deal with these issues will be examined. In this chapter, we will look at the following:

- Creating Java native OS threads
- Preventing race and deadlock conditions in threads
- Creating new virtual threads

Let us begin by looking at how we write threaded code using native threads.

Technical requirements

Here are the tools required to run the examples in this chapter:

- Java 17 installed to work with native threads only
- A text editor
- Maven 3.8.6 or a newer version installed

The sample code for this chapter is available at `https://github.com/PacktPublishing/Transitioning-to-Java/tree/chapter09`.

Creating Java native OS threads

The term *native thread* refers to threads managed by the computer's OS. When we create a Java native thread, we are referring to threads that the JVM manages using the underlying OS's threads library API. This also means that the JVM deals with the different threads libraries on different OSs, while we use the Java API to create threads. A program that employs threads written on an Apple Mac will work on a Windows machine as the JVM handles the lowest levels of threads.

We will look at three different ways to create Java native threads and one way to create a pool of threads. These will involve the following:

- Extending the `Thread` class
- Implementing the `Runnable` interface
- Creating a thread pool with `ExecutorService`
- Implementing the `Callable` interface
- Managing threads

The final items we will cover are the following:

- Daemon and non-daemon threads
- Thread priority

Extending the Thread class

Any class that extends the `Thread` class can contain methods that execute as part of a thread. Just creating an object does not create threads. Instead, a class that extends `Thread` must override the `run` method of the `Thread` superclass. Anything in this method becomes a thread. The `run` method can carry out all the work in the thread. Unless this is a simple task, the `run` method acts like the `perform` method I have used in my previous samples. Everything in the `run` method is what this thread will do. Let us look at a simple class that extends the `Thread` class.

We extend Thread to indicate that code in this class contains code that will be threaded:

```
public class ThreadClass extends Thread {
```

In this example, the work that each thread will do is count backward from whatever value we initially assign to a field called actionCounter:

```
private int actionCounter = 25;
```

Threads can be assigned a name. We want each thread in this example to have a number as its name. For this reason, it must be a static variable because there will just be one threadCounter integer, no matter how many instances of threadCounter we create. Static fields are considered thread-safe, meaning that there cannot be a conflict if two or more threads want access to the static field:

```
private static int threadCounter = 0;
```

The constructor is assigning the thread's name to the superclass's constructor. Each time we create an instance of this object, the value of threadCounter will be the same as what the previous instance set it as. This allows each thread to have a unique name:

```
public ThreadClass() {
    super("" + ++threadCounter);
}
```

The only task that this thread will perform is to display its name and the current value of the actionCounter field. We are overriding the toString method that is called whenever a reference to an object must act as String. It will return a string made up of the name assigned to it, by calling the superclass getName method, and the current value of actionCounter:

```
@Override
public String toString() {
    return "#" + getName() + " : " + actionCounter;
}
```

Thread classes must override the superclass run method. This is where the work of a thread happens. In this case, we are using an infinite while loop in which we are displaying this object's thread name and the current value of actionCounter. When actionCounter reaches zero, we return from the run method and the thread ends. The use of an infinite look syntax, while (true), means that the decision to end the loop is based on something that is happening in the loop, which, in this case, is decrementing actionCounter until it reaches zero. This is not the only way to write a run method, but it is the most common:

```
@Override
```

```
        public void run() {
            System.out.printf("extends Thread%n");
            while (true) {
                System.out.printf("%s%n", this);
                if (--actionCounter == 0) {
                    return;
                }
            }
        }
    }
```

With our threaded class in place, we can now write a class that will instantiate and run each of the threads:

```
public class ThreadClassRunner {
```

Here, in `perform`, we are instantiating five instances of `ThreadClass`. We call `start` and not `run`. The `start` method is an override of the `Thread` superclass's `start` method, and it is responsible for setting up the thread and then calling the `run` method. Calling the `run` method yourself will not start a thread:

```
        public void perform() {
            for (int i = 0; i < 5; i++) {
                new ThreadClass().start();
            }
        }
        public static void main(String[] args) {
            new ThreadClassRunner().perform();
        }
    }
```

> **Important note**
> Threads are not deterministic.

This is an important point to always be aware of. Each time you run this example code, the output will be different. The threads will execute in an order unrelated to the order they were created. Run this example a few times and take note that the order of the results is different every time.

There is one problem with this approach. You cannot extend any other superclass in `ThreadClass`. This brings us to the second approach.

Implementing the Runnable interface

In this approach, we implement the Runnable interface. We are performing the same task as in the previous example:

```
public class ThreadRunnableInterface implements Runnable{
    private int actionCounter = 25;
    @Override
    public String toString() {
```

We are calling Thread.currentThread().getName() to retrieve the name of this thread. When we extended the Thread class, we could call getName. As we are implementing the Runnable interface, we do not have superclass methods to call. We now get the name by using static methods of the Thread class, which will return information about the current thread that calls these methods:

```
        return "#" + Thread.currentThread().getName() +
                " : " + actionCounter;
    }
```

The run method is unchanged:

```
    @Override
    public void run() {
        while (true) {
            System.out.printf("%s%n", this);
            if (--actionCounter == 0) {
                return;
            }
        }
    }
}
```

The perform method in the class that starts the threads is different when using the Runnable interface:

```
    public void perform() {
        System.out.printf("implements Runnable%n");
        for (int i = 0; i < 5; i++) {
```

We create these threads by instantiating a Thread class, passing to its constructor an instance of the Runnable thread class along with the thread's name. In this example, the Thread object is anonymous; we do not assign it to a variable, and on it, we call start:

```
                        new Thread(new ThreadRunnableInterface(), ""
                                + ++i).start();
        }
    }
```

Just like in the previous example, the output will be different for each run.

Which technique should you use? The current best practice is to prefer the Runnable interface. This permits you to extend another class while it's still being threaded. Let us look at thread pooling.

Creating a thread pool with ExecutorService

What we have seen so far requires us to create a thread for every instance of the Thread class. An alternative approach is to create a pool of threads that can be reused. This is where the ExecutorService approach comes in. With this, we can create a pool of threads while, at the same time, defining the maximum number that can run concurrently. If more threads are required than the pool allows, then threads will wait until an executing thread ends. Let us change our basic example to use this service.

We begin with a class that implements Runnable. The actionCounter field is the number we will count down from in the thread:

```
public class ExecutorThreadingInterface implements Runnable {
    private int actionCounter = 250;
```

As we will leave the creation of the Thread class to ExecutorService, we no longer have a constructor that accepts String for the name of the thread. We will pass the name as int to the constructor and store it here. Fields in a class that becomes a single thread will each have their own actionCounter and threadCount instances:

```
    private final int threadCount;
```

Here is the constructor that takes the name we want to know the thread by:

```
    public ExecutorThreadingInterface(int count) {
        threadCount = count;
    }
```

We are overriding the toString method to return String with the current thread's name, which is assigned by ExecutorService, along with the name we assigned to threadCount as a number, followed by the current value of actionCounter, which decreases while the thread runs:

```
    @Override
    public String toString() {
```

```
            return "#" + Thread.currentThread().getName()
                + "-" + threadCount + " : " + actionCounter;
        }
```

The last method is run. This remains unchanged:

```
        @Override
        public void run() {
            while (true) {
                System.out.printf("%s%n", this);
                if (--actionCounter == 0) {
                    return;
                }
            }
        }
    }
```

Now, let us look at how we use `ExecutorService` to create the threads:

```
    public class ExecutorServiceRunner {
```

I have chosen to make the variables we use as fields. They could all just be declared as local variables in a single method:

```
        private final int numOfThreads = 5;
        private final int threadPoolSize = 2;
        private final ExecutorService service;
```

The constructor now has the responsibility to instantiate `ExecutorService`, along with an array of `Runnable` threads:

```
        public ExecutorServiceRunner() {
            service =
                Executors.newFixedThreadPool(threadPoolSize);
        }
        public void perform() {
            for (int i = 0; i < numOfThreads; i++) {
```

We add threads to `ExecutorService` using the `execute` method. We do not need access to the threads, so they are instantiated anonymously:

```
                    service.execute(
                            new ExecutorThreadingInterface(i));
        }
```

The service will shut down after all threads have finished. You can no longer add any threads to the service after this method is called:

```
            service.shutdown();
        }
```

We end this class with the usual main method:

```
        public static void main(String[] args) {
            new ExecutorServiceRunner().perform();
        }
    }
```

These three approaches allow you to easily create threads. One issue that they all share is that when the thread ends, it does not return a value because run is void. We can resolve this, if needed, by combining ExecutorService with a third type of threaded class by using the Callable interface.

Implementing the Callable interface

In each of the thread classes we have seen, they have all had a run method that returned void when the thread ended. This leads us to the Callable interface. Using this interface, the end of a thread returns a value. We can only use this technique if we use ExecutorService. Let's begin by looking at a Callable thread class.

We begin with the class that we want to thread. We implement the Callable interface, and using generic notation, we state that the value returned when the thread ends will be a string. The fields, constructor, and toString are the same as ExecutorThreadingInterface:

```
public class ThreadCallableInterface
                            implements Callable<String> {
    private int actionCounter = 250;
    private final int threadCount;

    public ThreadCallableInterface(int count) {
        threadCount = count;
    }
    @Override
```

```java
    public String toString() {
        return "#" + Thread.currentThread().getName() +
                "-" + threadCount + " : " + actionCounter;
    }
}
```

Here, we replace `run` with `call` and show a return type. The `return` statement will display the thread name that we assigned as an integer to each thread.

```java
    @Override
    public String call() {
        while (true) {
            System.out.printf("%s%n", this);
            if (--actionCounter == 0) {
                return "Thread # " + threadCount +
                                        " is finished";
            }
        }
    }
}
```

Now, let us look at the runner for this `Callable` thread.

```java
public class ThreadCallableInterfaceRunner {
```

The first variable we are declaring is `List` of the `Future` type. `Future` is an interface, like `List`. When we use, not execute, the `submit` method of `ExecutorService`, it returns an object that implements the `Future` interface. Objects that implement this interface represent the result of an asynchronous task. When we instantiate this object in a few lines from here, it will be `List` of `Future` strings that are delivered by the thread:

```java
    private final List<Future<String>> futureList;
    private final ExecutorService executor;

    private final int numOfThreads = 5;
    private final int threadPoolSize = 2;
```

A new feature in this example is the display of the current date and time that each thread ends. The `DateTimeFormatter` object converts an `LocalDateTime` object into a readable string:

```java
    private final DateTimeFormatter dtf;
```

The constructor instantiates the class fields:

```
public ThreadCallableInterfaceRunner() {
    executor =
        Executors.newFixedThreadPool(threadPoolSize);
```

We instantiate `futureList` as `ArrayList`. We follow this by defining the format of the date and time we want:

```
    futureList = new ArrayList<>();
    dtf = DateTimeFormatter.ofPattern(
                        "yyyy/MM/dd HH:mm:ss");
}
public void perform(){
```

Here, we use `submit` to submit the threads to the `ExecutorService`. Using `submit` implies that we expect a return of type `Future`. We also add each `Future` object to an `ArrayList` instance:

```
for (int i = 0; i < numOfThreads; i++) {
    Future<String> future = executor.submit(
                    new ThreadCallableInterface(i));
    futureList.add(future);
}
```

Here, we cycle through the `ArrayList` instance displaying the current date and time along with the thread's returned value – in this example, `String`. We access the return value of the `Future` object by calling the `get` method. This is a blocking method call. Each `Future` object is associated with a specific thread, and `get` will wait for its result before it allows the next `Future` object's `get` to execute. A call to `get` can result in two checked exceptions, so we must place the call in a `try/catch` block. For the purpose of this example, we are just printing the stack trace. You should never just print a stack trace without taking any appropriate action:

```
for (Future<String> futureResult : futureList) {
    try {
        System.out.println(
            dtf.format(LocalDateTime.now()) + ":" +
            futureResult.get());
    } catch (InterruptedException |
                    ExecutionException e) {
        e.printStackTrace();
```

```
            }
        }
```

`ExecutorService` must be explicitly shut down when you no longer need the service:

```
        executor.shutdown();
    }
```

We end this with the usual `main` method:

```
    public static void main(String[] args) {
        new ThreadCallableInterfaceRunner().perform();
    }
}
```

Now that we have reviewed the most common approaches to creating `Threads`, let us look a little deeper at how we can manage a thread.

Managing threads

There are three `Thread` methods that are commonly used to manage a thread. These are as follows:

- `yield()`
- `join()`
- `sleep()`

The `yield()` method informs the thread scheduler that it can give up its current usage of the processor but wishes to be rescheduled as soon as possible. This is only a suggestion, and the scheduler is free to do what it wishes. This makes it non-deterministic as well as dependent on the platform it is running on. It should only be used when it can be proved, usually by profiling the code, that it can improve performance.

The `join()` method can be useful when one thread (we will call it **A**), starts another thread (we will call it **B**). If we require the **B** thread to complete its task before the **A** thread, then we can use a join on the **A** thread, and that will block the **A** thread until **B** finishes its task. Using `join()` affects the first thread that created the second thread. There are two additional overloaded versions of join that allow you to set the length of time to block the thread that started it in either milliseconds, or milliseconds and nanoseconds.

The final method is a static method of the `Thread` class. The `sleep()` method pauses the thread it is executed in for a specific length of time. The time can be, like join, in milliseconds, or milliseconds and nanoseconds.

One characteristic of both `join()` and `sleep()` is that they can throw checked exceptions. They must be coded inside a `try/catch` block. The following is a thread class that instantiates and starts a second class but then joins the second class, thus blocking itself until the thread it started finishes.

This is like the first `ThreadClass` instance we had seen. The difference is that it instantiates another thread class and then starts that thread at the beginning of the `run` method:

```java
public class ThreadClass1 extends Thread {
    private int actionCounter = 500;
    private static int threadCounter = 0;
    private final ThreadClass2 tc2;

    public ThreadClass1() {
        super("" + ++threadCounter);
        tc2 = new ThreadClass2();
    }
    @Override
    public String toString() {
        return "#" + getName() + " : " + actionCounter;
    }
    @Override
    public void run() {
```

Here, we start the second thread, and it will now execute as determined by the scheduler:

```java
        tc2.start();
        while (true) {
            System.out.printf("%s%n", this);
```

When the first thread reaches 225, we issue a join on the second thread. The result will be that the first thread is blocked, and the second thread will run till it finishes before unblocking the first thread:

```java
            if (actionCounter == 225) {
                try {
                    tc2.join();
                } catch (InterruptedException ex) {
                    ex.printStackTrace();
                }
            }
            if (--actionCounter == 0) {
```

```
                    return;
            }
        }
    }
}
```

Daemon and non-daemon threads

The term *daemon* refers to what is considered a low-priority thread. What this means is that any such native threads designated as daemons will end, regardless of what they are doing at that moment in time when the application's main thread ends.

A non-daemon thread, the default for when a native thread is created, will block the application's main thread from ending until the thread completes its task. This tells us that non-daemon threads must have an ending condition. A daemon thread does not need an ending condition, as it will end with the main thread.

You can set a native thread to be a daemon with a simple method call:

```
thread.setDaemon(true);
```

You can only call this method on a thread after it has been instantiated but before it has started. You cannot change the daemon status after it starts. Calling this method will result in an exception being thrown.

Thread priority

As already pointed out, threads are non-deterministic. This means we have no absolute control over when a thread will get its slice of time to run or how long that time slice will be. We can make a suggestion, also called a hint, and that is where thread priority comes in.

The range of possible values for thread priority is 1 through 10. There are three defined static constants that are used in most cases rather than a number. These are as follows:

- `Thread.MAX_PRIORITY`
- `Thread.MIN_PRIORITY`
- `Thread.NORM_PRIORITY`

As this suggests, you cannot rely on maximum priority getting more time slices than minimum priority. It should get more slices, and that is the best that you can hope for.

Now that we have seen how to manage our threads, let us look at one more topic, thread safety.

Preventing race and deadlock conditions in threads

There are two common problems that can cause problems with threaded code. The first is the race condition. This is what can happen when two or more threads work with a block of code that changes a variable shared by all the threads.

The second is the deadlock condition. To resolve race conditions, you lock a block of code. If multiple threads use the same lock object, then you could have a situation where these threads are waiting for each other to finish with the lock but none finish. Let us look more closely at these two conditions.

Race condition

Imagine a scenario where you share a reference to an object among multiple threads. Calling upon methods in this shared class that only use local variables is thread-safe. Thread-safe, in this case, occurs because each thread maintains its own private stack for local variables. There can be no conflict between threads.

It is a different story if the shared object's methods access and alter a class field. Unlike local variables, class fields are unique to the shared object and every thread that calls methods in this class is possibly altering a field. Operations on these fields may not finish before the thread's time slice ends. Now, imagine that a thread expects a field to have a specific value based on the last time slice that accessed the field. Unbeknown to it, another thread has changed that value. This results in what is referred to as a race condition. Let us look at an example.

Here is a simple class that adds a passed value to a field of the class called `counter` and returns the result of adding the value. Every time we call `addUp`, we expect the `counter` field to change the value:

```java
public class Adder {
    private long counter = 0;
    public long addUp(long value) {
        counter += value;
```

How much time a thread gets from the scheduler is related to the CPU of your computer. A high clock rate along with multiple CPU cores sometimes permits a thread to complete its task before the next thread takes over. For that reason, I have slowed down the `addUp` method by having it sleep for a half second:

```java
        try {
            Thread.sleep(500);
        } catch (InterruptedException ex) {
            ex.printStackTrace();
        }
        return counter;
```

```
    }
}
```

The threaded class is based on `ThreadClass` we have already seen:

```
public class SynchronizedThreadClass extends Thread {

    private int actionCounter = 5;
    private static int threadCounter = 0;
```

We have a field for holding a reference to an `Adder` object. No matter how many threads we create, they will all share the field variables:

```
    private final Adder adder;
```

The constructor receives a reference to the `Adder` object:

```
    public SynchronizedThreadClass(Adder adder) {
        super("" + ++threadCounter);
        this.adder = adder;
    }
    @Override
    public String toString() {
        return "#" + getName() + " : " + actionCounter;
    }
    @Override
    public void run() {
        while (true) {
            var value = adder.addUp(2);
```

Here, we are printing information on this thread and the current value from the `addUp` method:

```
            System.out.printf(
                "%s : %d%n", this, adder.addUp(2));
            if (--actionCounter == 0) {
                return;
            }
        }
    }
}
```

Here is the main class:

```
public class SynchronizedExample {
```

We will have two threads of type SynchronizedThreadClass:

```
    private final SynchronizedThreadClass tc1;
    private final SynchronizedThreadClass tc2;
    private final Adder sa;
```

Before we instantiate each thread, we create a single Adder object that we share with each threaded class:

```
    public SynchronizedExample() {
        sa = new Adder();
        tc1 = new SynchronizedThreadClass(sa);
        tc2 = new SynchronizedThreadClass(sa);
    }
    public void perform() {
        tc1.start();
        tc2.start();
    }
    public static void main(String[] args) {
        new SynchronizedExample().perform();
    }
}
```

This code is not yet synchronized. Here is a table of results when access to the adder is not synchronized:

Unsynchronized		
Thread	Thread class actionCounter	Adder class counter
#1	5	4
#2	5	4
#1	4	8
#2	4	10
#1	3	12

#2	3	14
#1	2	16
#2	2	18
#1	1	20
#2	1	20

Table 9.1 – The results from running the code unsynchronized

The expectation is that the Adder class counter should count from 2 to 20. It does not. The first thread begins adding the value passed, 2, to the counter. But before it can display its result, the second thread comes along and adds 2 to the same counter, now increasing the value to 4. When we return to the first thread that will just display the result, it is now 4, and not the value of 2 that it had when its first time slice ended. If you run this multiple times, the results will be different, but we will see this problem in other places in the output.

Now, let us synchronize the code. Synchronizing applies a lock to a section of code commonly called a critical section. The lock is a reference to an object, as all objects, by virtue of their Object superclass, can be used as a lock. We only need to change the Adder class and specifically the addUp method:

```
public long addUp(long value) {
    synchronized (this) {
        counter += value;
        try {
            Thread.sleep(10);
        } catch (InterruptedException ex) {
            ex.printStackTrace();
        }
        return counter;
    }
}
```

As the entire method is considered a critical section, we can remove the synchronize block and apply the synchronized keyword to the method name:

```
public synchronized long addUp(long value) {
```

Here is the table of results using the synchronized version of the addUp Adder class method:

Synchronized		
Thread	Thread class actionCounter	Adder class counter
#1	5	2
#2	5	4
#1	4	6
#2	4	8
#1	3	10
#2	3	12
#1	2	14
#2	2	16
#1	1	18
#2	1	20

Table 9.2 – The results from running the code synchronized

You can see that every value from 2 to 20 appears.

As a developer, you are always looking for tasks that can be carried out concurrently. Once identified, you will apply threading where appropriate. Any long-running task is a candidate for a thread. User interfaces typically run the interface in one or more threads, and as tasks are selected from menus or buttons, they too are run in threads. This means that the user interface can respond to you even while it is performing a long-running task.

Now, let us look at a problem that can arise if we synchronize blocks of code with the same lock object improperly.

Deadlock condition

A thread deadlock occurs when thread locks get intertwined, especially if a thread is nested inside another one. This results in each thread waiting for the other to end. When using synchronize, a lock can be any object or class in Java that will protect a critical section, usually to avoid the race condition. You can also create objects of type Lock or ReentrantLock. Either approach, as we shall see, can result in a deadlock. Deadlock can be difficult to recognize, as it does not crash the program or throw an exception. Let us look at an example of code that will be deadlocked.

We begin with the class that will create the lock objects and then we start two threads with them:

```
public class Deadlock1 {
```

Here are the two lock objects we will use with synchronize in Thread1 and Thread2. Any object in Java, either one you create or one that already exists, such as String, can be used as a lock:

```
public final Object lock1 = new Object();
public final Object lock2 = new Object();

public void perform() {
    var t1 = new ThreadLock1(lock1, lock2);
    var t2 = new ThreadLock2(lock1, lock2);
    t1.start();
    t2.start();
}

public static void main(String args[]) {
    new Deadlock1().perform();
}
}
```

Now, let us look at the class that extends Thread. Take note of the fact that Thread1 uses lock1 before lock2, while Thread2 uses lock2 before lock1:

```
public class ThreadLock1 extends Thread {
    private final Object lock1;
    private final Object lock2;
    public ThreadLock1(Object lock1, Object lock2) {
        this.lock1 = lock1;
        this.lock2 = lock2;
    }
}
```

In this run method, we have a synchronized block that uses lock1 and then a nested synchronized block that uses lock2:

```
@Override
public void run() {
    synchronized (lock1) {
        System.out.printf(
                "Thread 1: Holding lock 1%n");
        try {
            Thread.sleep(10);
```

```
            } catch (InterruptedException e) {
            }
            System.out.printf(
                    "Thread 1: Waiting for lock 2%n");
            synchronized (lock2) {
                System.out.printf(
                        "Thread 1: Holding lock 1 & 2%n");
            }
        }
    }
}

public class ThreadLock2 extends Thread {
    private final Object lock1;
    private final Object lock2;
    public ThreadLock2(Object lock1, Object lock2) {
        this.lock1 = lock1;
        this.lock2 = lock2;
    }
```

In this run method, we have a synchronized block that uses lock2 and then a nested synchronized block that uses lock1:

```
    @Override
    public void run() {
        synchronized (lock2) {
            System.out.printf(
                    "Thread 2: Holding lock 2%n");
            try {
                Thread.sleep(10);
            } catch (InterruptedException e) {
            }
            System.out.printf(
                    "Thread 2: Waiting for lock 1%n");
            synchronized (lock1) {
                System.out.printf(
                        "Thread 2: Holding lock 1 & 2%n");
```

```
                    }
                }
            }
        }
```

When we run this code, the output will be as follows:

```
Thread 1: Holding lock 1
Thread 2: Holding lock 2
Thread 1: Waiting for lock 2
Thread 2: Waiting for lock 1
```

The program is now deadlocked, and both threads are waiting for the other thread to finish. If we change the order of the locks we use so that both use lock1 first and lock2 second, we will get the following:

```
Thread 1: Holding lock 1
Thread 1: Waiting for lock 2
Thread 1: Holding lock 1 & 2
Thread 2: Holding lock 2
Thread 2: Waiting for lock 1
Thread 2: Holding lock 1 & 2
```

The deadlock condition is solved. Deadlocks are rarely this obvious, and you may not even be aware that there is a deadlock happening. You need a thread dump to determine whether there is a deadlock in your code.

Here's one final point on this topic – rather than using an instance of the Object class as the lock, you can use the Lock class. The syntax is a little different, and you can ask a Lock object whether it is being used. The following code snippet shows what it will look like, but it does not solve the deadlock.

In the main class, the locks will use the Lock interface with the ReentrantLock class that implements the interface:

```
        public final Lock lock1 = new ReentrantLock();
        public final Lock lock2 = new ReentrantLock();
```

In this code, we are passing the Lock objects to the class through the constructor:

```
public class ThreadLock1a extends Thread {
    private final Lock lock1;
    private final Lock lock2;
```

```
public ThreadLock1a(Lock lock1, Lock lock2) {
    this.lock1 = lock1;
    this.lock2 = lock2;
}
@Override
public void run() {
```

Note that we do not use a synchronized block, but instead, we call lock on lock1, and when the critical section is finished, we issue unlock on lock1:

```
lock1.lock();
System.out.printf("Thread 1a: Holding lock 1%n");
try {
    Thread.sleep(10);
} catch (InterruptedException e) {
}
System.out.printf(
        "Thread 1a: Waiting for lock 2%n");
lock2.lock();
System.out.printf("Thread 1a: Holding lock 1 & 2");
lock2.unlock();
lock1.unlock();
    }
}
```

One advantage of using lock objects rather than a synchronized block is that the call to unlock does not have to be in the same class.

Multithreading is a powerful feature of Java that you should use when appropriate. What we saw is based on native threading available in Java from version 1.0. Recently, a new type of threading was introduced. Let us look at it.

Creating new virtual threads

As pointed out at the beginning of the previous section, native Java threads are managed by the JVM by working directly with the OS's threading library. There is a one-to-one relationship between a native thread and an OS thread. This new approach is called **virtual threads**. While native threads are managed by the JVM in collaboration with the OS, virtual threads are managed exclusively by the JVM. OS threads are still used, but what makes this approach significant is that virtual threads can share OS threads and are no longer one-to-one.

Virtual threads do not run faster and can suffer from race and deadlock conditions. What is special with virtual threads is that the number of threads that you can start up could be in the millions. How we use virtual threads is not much different from native threads. The following code snippet shows the `perform` method that we saw in the previous examples creating virtual threads. The thread class is unchanged, thus making the use of a virtual thread rather than a native thread quite easy:

```
public void perform() {
    for (int i = 0; i < 5; ++i) {
```

Here, we are creating a virtual thread and starting it:

```
        Thread.ofVirtual().name("Thread # " + i).
            start(new VirtualThreadRunnableInterface());
    }
    try {
        Thread.sleep(500);
    } catch (InterruptedException ex) {
        ex.printStackTrace();
    }
}
```

I must embarrassingly admit that it took me 3 days to get this code to work. Why? I neglected an important characteristic of virtual threads – they are daemon threads. Attempting to make them non-daemon has no effect. What was happening to me is that the program would end before any output appeared, or only some appeared but not all of the expected output from the threads. When `perform` ended and returned to the `main()` method, the main native thread ended. When this happened, all daemon threads were ended. My PC executed `perform`, returned to `main`, and ended before a single virtual thread could display its output.

You can see the solution I used to make this code work. I called `Thread.sleep()`. This puts the current thread to sleep for a specified length of time. In this case, 500 milliseconds was enough time for the virtual threads to complete all their tasks before the main thread ended.

Finally, you cannot change the priority of virtual threads. They all run at NORM_PRIORITY.

Summary

As mentioned at the start of this chapter, Java's native support of threads is one of the reasons for its popularity. In this chapter, we saw how we can create threads by extending the `Thread` class and by implementing either a `Runnable` or `Callable` interface. We saw how `ExecutorService` allows us to pool threads. We concluded by looking at one specific issue, where two or more threads compete for access to a shared resource, called a race condition, and saw how we resolve this by applying locks through the application of synchronization.

There are changes coming to threading. Project Loom, at the time of writing, introduces threads managed exclusively by the JVM along with a framework for concurrency. Some features are in preview, while others are in incubation. It will be a few years before these new types of threads become commonplace. I recommend following the development of this project.

In our next chapter, we will look at the most used design patterns in Java development. These patterns will provide us with well-established approaches to organizing our code.

Further reading

- *Creating and Starting Java Threads*: `https://jenkov.com/tutorials/java-concurrency/creating-and-starting-threads.html`

- *Introduction to Thread Pools in Java*: `https://www.baeldung.com/thread-pool-java-and-guava`

- *Deadlock in Java Multithreading*: `https://www.geeksforgeeks.org/deadlock-in-java-multithreading/`

- Here is a book from 2006, that remains one of the finest references on threading in Java: *Java Concurrency in Practice, 1st Edition*, by Brian Goetz with Tim Peierls, Joshua Bloch, Joseph Bowbeer, David Holmes, and Doug Lea (ISBN-13: 978-0321349606)

- *JEP 425: Virtual Threads*: `https://openjdk.org/jeps/425`

10
Implementing Software Design Principles and Patterns in Java

Software design principles provide guidance on how you construct your classes and how your objects should interact. They are not tied to a specific problem. For example, the **single responsibility** principle encourages us to write methods that perform a single task. Software design patterns are reusable concepts for solving common problems in software design. For example, should we need to have a single instance of an object in an application, you will want to use the **Singleton** pattern. This pattern has nothing to do with the language you are using, nor does it describe the required code for the pattern. What these principles and patterns do is describe a solution to common problems that you can then implement in the language you are using. Principles and patterns can be applied to any language, and I will assume that you have likely applied them in whatever language you are coming from.

The goal of this chapter is to look at a handful of the most commonly used principles and patterns and how they are coded in Java. They are as follows:

- SOLID software design principles
- Software design patterns

As a software developer, you are expected to write code that is reusable, understandable, flexible, and maintainable.

Technical requirements

Here are the tools required to run the examples in this chapter:

- Java 17 installed
- Text editor
- Maven 3.8.6 or a newer version installed

The sample code for this chapter is available at `https://github.com/PacktPublishing/Transitioning-to-Java/tree/chapter10`.

SOLID software design principles

Software design principles, as applied to object-oriented programming, provide guidelines for how you construct your classes. Unlike patterns that are tied to specific coding requirements, principles should be considered for any code that you write every day. There are numerous principles, but we will look at five principles that fall under the acronym **SOLID**.

S – Separation of concerns/single responsibility

In my opinion, **separation of concerns** may be the most important of the principles. Simply put, it directs us to design classes that are responsible for a specific function in the program. We have seen this already in *Chapter 5, Language Fundamentals – Classes*, in the *Class organization based on functionality* section, where we took the CompoundInterest05 program and organized the classes based on functionality:

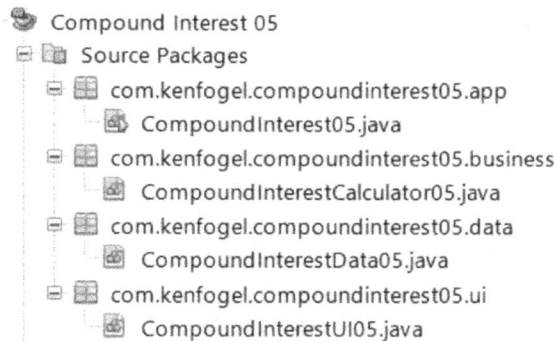

Figure 10.1 – Class organization for the separation of concerns principle

What we see in this figure are four packages that describe the required functionality of any classes in the package. The individual classes in each package contain any methods and data structures required to carry out their task. The business class is independent of the **user interface** (**UI**) class. We can change the UI class to employ a GUI rather than a text or console interface. We can do this without the need to make any changes to the business class. We will do just this in *Chapter 13, Desktop Graphical User Interface Coding with Swing and JavaFX*, when we change the UI from text to GUI. We will even turn this application into a web app in *Chapter 15, Jakarta Faces Application*, leaving the business and data classes untouched.

The single responsibility principle is not technically part of SOLID but its close association with the separation of concerns merits including it here. This principle is applied to methods. A single method should be written to be responsible for a single concern. You would not write a single method that performs a calculation and then displays the result. These are two concerns, and each belongs to its own method.

O – Open/closed

This principle states that you should be able to add to or extend a class's functionality, the open part, without changing or modifying any code in the class, the closed part. This can be accomplished using either inheritance or interface. Consider our CompoundInterestCalculator05.java example. It contains a single calculation for compound interest. What if I'd like to add another calculation? Should I just edit this file and add a new method for the new calculation? The answer is no.

If I use inheritance, I will create a new class that extends the original single calculation and adds the new calculation. Here is a class that implements a method to calculate loan payments:

```java
public class Calculation {
    public void loanCalculator(FinancialData data) {
        var monthlyPayment = data.getPrincipalAmount() *
            (data.getMonthlyInterestRate() / (1 - Math.pow(
            (1 + data.getMonthlyInterestRate())), -
            data.getMonthlyPeriods()))));
        data.setMonthlyPayment(monthlyPayment);
    }
}
```

You have thoroughly tested this calculation and you are confident that it is correct. Now, you are asked to add two more calculations, one for the savings goal and the other for the future value of regular savings. The open/closed principle tells us that we should not modify this class. Although it's a simple example, there is still the possibility that you may inadvertently change something in this class when you add methods to it. The solution is inheritance:

```java
public class Calculation2 extends Calculation {
    public void futureValueCalculator(FinancialData data) {
        var futureValue = data.getMonthlyPayment() *
            ((1 - Math.pow(
            (1 + data.getMonthlyInterestRate()),
            data.getMonthlyPeriods())) /
            data.getMonthlyInterestRate());
        data.setPrincipalAmount(futureValue);
    }
}
```

```java
        public void savingsGoalCalculator(FinancialData data) {
            double monthlyPayment = data.getPrincipalAmount() *
                (data.getMonthlyInterestRate() /
                (1 - Math.pow(
                (1 + data.getMonthlyInterestRate()),
                data.getMonthlyPeriods()))));
            data.setMonthlyPayment(monthlyPayment);
        }
    }
```

This `Calculator2` class inherits the public `loanCalculator` method from the `Calculator` superclass and then adds the two new calculations.

The second approach, called the polymorphic open/closed principle, is to use an interface class. This is the closed class. All calculations must implement this interface:

```java
public interface FinanceCalculate {
    void determine(FinancialData data);
}
```

Now, let us look at one of the three classes that will implement this interface:

```java
public class FutureValue implements FinanceCalculate {
    @Override
    public void determine(FinancialData data) {
        var futureValue = data.getMonthlyPayment() *
            ((1 - Math.pow(
            (1 + data.getMonthlyInterestRate()),
            data.getMonthlyPeriods())) /
            data.getMonthlyInterestRate());
        data.setPrincipalAmount(futureValue);
    }
}
```

Now, we can write a class that can call upon any of these operations. If we wish to add new financial calculations, we can without modifying this class, as it expects to receive a reference to an object that implements the `FinanceCalculate` interface:

```java
public class BankingServices {
    public void doCalculation(FinanceCalculate process,
```

```
                                   FinancialData data) {
        process.determine(data);
    }
}
```

As new features are added to an application, you want to accomplish this without modifying existing code. There is one exception, and that is to correct bugs. This will likely require modifications to the code.

L – Liskov substitution

This principle describes how to use inheritance effectively. Put simply, when you create a new class that extends an existing class, you can override methods in the superclass. What you must not do is reuse a superclass method name and change its return type or the number or type of parameters.

Here is a very simple superclass that displays a message:

```
public class SuperClass {
    public void display(String name) {
        System.out.printf("Welcome %s%n", name);
    }
}
```

Now, let us create a subclass that displays a slightly different message:

```
public class SubClass extends SuperClass {
    @Override
    public void display(String name) {
        System.out.printf("Welcome to Java %s%n", name);
    }
}
```

When we override the display method, we are not changing its return value or its parameters. This means that, in the following code, we can use either the superclass or the subclass, and the version matching the reference type passed to doDisplay will run:

```
public class Liskov {
    public void doDisplay(SuperClass sc) {
        sc.display("Ken");
    }
    public static void main(String[] args) {
        new Liskov().doDisplay(new SuperClass());
```

```
        }
    }
```

The program's output will be as follows:

Welcome Ken

Now, let us pass a reference to SubClass:

```
            new Liskov().doDisplay(new SubClass());
```

The output will now be as follows:

Welcome to Java Ken

Changing the return type of display in the subclass is a compiler error. Leaving the return type as void but adding or removing parameters breaks the override. The superclass display method is the only one that can be called if the display method's reference is to SubClass. Here is the new SubClass:

```
public class SubClass extends SuperClass {
    public void display(String name, int age) {
        System.out.printf("Welcome to Java %s at age %d%n",
                          name, age);
    }
}
```

We cannot use the @Override annotation as this is considered overloading, keeping the same method name but changing parameters. Now, if we pass SubClass to doDisplay, the method chosen will always be the SuperClass version, thus breaking the Liskov principle.

I – Interface segregation

This principle provides guidance on developing interfaces. Simply put, do not add new methods to an interface that not every implementation of an interface will require. Let us look at a simple interface for a delivery service. Keep in mind that, for brevity, this and many other examples are not complete but show what is relevant to the concept being explained:

```
public interface Delivery {
    void doPackageSize(int length, int height, int width);
    void doDeliveryCharge();
}
```

Now, let us implement this interface:

```java
public class Courier implements Delivery {
    private double packageSize;
    private double charge;
    @Override
    public void doPackageSize(int length, int height,
                                               int width) {
        packageSize = length * width * width;
    }
    @Override
    public void doDeliveryCharge() {
        if (packageSize < 5) {
            charge = 2.0;
        } else if (packageSize < 10 ) {
            charge = 4.0;
        } else {
            charge = 10.0;             }
    }
}
```

Now, imagine that we need to expand `Courier` to handle packages that will travel by air freight. We now must add methods just used for this type of transport. Do we add this to the existing interface? No, we do not. Interface segregation tells us to keep interfaces to the minimum number of methods required for a specific use. The new interface could look like this:

```java
public interface AirDelivery extends Delivery {
    boolean isHazardous();
}
```

We are using interface inheritance here. With this interface, you will need to implement the methods from `Delivery` as well as the new method in `AirDelivery`. Now, if we implement a class with the `Delivery` interface, we only need to implement two methods. When we use `AirDelivery`, we need to implement three methods.

D – Dependency inversion

This last principle in SOLID states that classes should not depend on a concrete class. Rather, classes should depend on abstraction. An abstraction can be an abstract class or, more commonly in Java, an interface. Imagine a program that deals with the inventory for a store. We would create a class for each item. We could have the following:

```java
public class Bread {
    private String description;
    private int stockAmount;
    public String getDescription() {
        return description;
    }
    public void setDescription(String description) {
        this.description = description;
    }
    public int getStockAmount() {
        return stockAmount;
    }
    public void setStockAmount(int stockAmount) {
        this.stockAmount = stockAmount;
    }
}
```

For another item, such as `Milk`, we could have the following:

```java
public class Milk {
    private String description;
    private int stockAmount;
    public String getDescription() {
        return description;
    }
    public void setDescription(String description) {
        this.description = description;
    }
    public int getStockAmount() {
        return stockAmount;
    }
    public void setStockAmount(int stockAmount) {
        this.stockAmount = stockAmount;
    }
}
```

A program that uses either `Bread` or `Milk` is considered a high-level module, while `Bread` and `Milk` are considered concrete low-level modules. It also implies that the high-level module must depend on the concrete class. Imagine that we need a program to produce a report on items in inventory. Without following dependency inversion, we will need one report class for every item in the inventory:

```
public class MilkReport {
    private final Milk milkData;
    public MilkReport(Milk data) {
        milkData = data;
    }
    public void displayReport() {
        System.out.printf("Description: %s  Stock: %d%n",
            milkData.getDescription(),
            milkData.getStockAmount());
    }
}
```

We will now need a second class for `BreadReport`. A store with 100 items for sale will need 100 classes, one for each item. The problem that dependency inversion resolves is the need for 100 report classes. We begin the solution using an interface:

```
public interface Inventory {
    public String getDescription();
    public void setDescription(String description);
    public int getStockAmount();
    public void setStockAmount(int stockAmount);
}
```

Now, every item class will implement `Inventory`:

```
public class MilkDI implements Inventory{ . . . }
public class BreadDI implements Inventory{ . . . }
```

There can now be just one report class:

```
public class InventoryReport {
    private final Inventory inventoryData;
    public InventoryReport(Inventory data) {
        inventoryData = data;
    }
```

```
    public void displayReport() {
        System.out.printf("Description: %s   Stock: %d%n",
            inventoryData.getDescription(),
            inventoryData.getStockAmount() );
    }
}
```

In using dependency inversion, your program can eliminate redundancy while still permitting your program to process any new items in the inventory.

Software design principles will contribute to writing efficient and maintainable code. These principles, along with other principles, should be considered every time you write a line of code. You can read more about these principles and others from links in the *Further reading* section.

Principles should govern every line of code that you write. The upcoming patterns guide you in solving specific problems.

Software design patterns

Software design patterns describe solutions to specific issues in software. This concept comes from architecture and engineering. Imagine that you need to design a bridge to cross a river. You will likely begin by choosing the bridge type or pattern. There are seven types of bridges:

- Arch bridge
- Beam bridge
- Cantilever bridge
- Suspension bridge
- Cable-stayed bridge
- Tied-arch bridge
- Truss bridge

These types, or patterns, describe how the bridge should span the river you wish to build your bridge over, but they do not provide detailed instructions or blueprints. They guide the architect in the design of the bridge. Software patterns work in a similar fashion. Let us look at four widely used patterns and how they could be implemented in Java.

Singleton

A singleton is a Java object that can only be instantiated once. It is a **creational** pattern. Wherever this object is used in an application, it is always the same object. In an application that requires exclusive access to resources by means of passing a token object, a singleton is one pattern you could follow. In *Chapter 11, Documentation and Logging*, we will look at logging, and most logging frameworks use a singleton logger object. Otherwise, there would be a separate logger in every class that uses it. Objects that manage thread pools are also frequently written as singletons.

Implementing a singleton in Java can be quite easy. In the examples that follow, the singletons are doing nothing other than ensuring that there will be only one instance of them. I leave it to you to add the actual work that these singletons should perform.

```
public class SingletonSafe {
```

We use a static variable to represent this object when the `getInstance` method is called. As this method is static, it can only access static fields in the class. Static fields are also shared by all instances of an object.

```
    private static Singleton instance;
```

Up until now, all constructors were `public`. Designating the constructor as `private` means that you cannot instantiate this object with `new`.

```
    private Singleton() {}
```

In this method, we test to see whether an instance already exists. If it does, then we return the instance. If it does not, then we instantiate the object with `new`. But wait, I just wrote that you cannot use `new` on a class with a `private` constructor. This is true if this object is instantiated in another object. Here, we are instantiating the object inside itself, and access control does not apply. Every method in a class can access any other method regardless of its access. So, while the constructor is `private` to the `getInstance` method, it can be run when the object is created with `new`:

```
    public static Singleton getInstance() {
        if (instance == null) {
            instance = new Singleton();
        }
        return instance;
    }
}
```

Now, let us test whether this is working:

```
public class SingletonExample {

    public void perform() {
```

Here, we are instantiating two class fields by calling upon the getInstance method of Singleton:

```
var myInstance1 = Singleton.getInstance();
var myInstance2 = Singleton.getInstance();
```

If our Singleton class is working, the two instances of Singleton will be the same. When we compare object references, we are comparing the address in memory of these objects. If the address is the same, then we have the same object:

```
if (myInstance1 == myInstance2) {
    System.out.printf(
        "Objects are the same%n");
} else {
    System.out.printf(
        " Objects are different%n");
    }
}
public static void main(String[] args) {
    new SingletonExample().perform ();
    }
}
```

There is one problem with our Singleton class. It is not thread-safe. It is possible that getInstance could be interrupted by a thread or threads, and this could result in two or more instances of Singleton. We can make this class thread-safe by synchronizing the creation of the object. Here is the updated getInstance method:

```
public static Singleton getInstance() {
```

When we create a synchronized block, we are guaranteeing that the instantiation of the Singleton class cannot be interrupted. This ensures that all threads get the same instance after the one time it is instantiated:

```
synchronized (Singleton.class) {
    if (instance == null) {
        instance = new Singleton();
    }
}
return instance;
}
```

With this, we now have a thread-safe Singleton class.

Factory

A factory, another creational pattern, is a class that instantiates a specific class from a family of classes that either share the same interface or are all subclasses of the same abstract class. It is a **creational** pattern. We will look at an example that uses an interface. Here is the shared interface:

```
public interface SharedInterface {
    String whatAmI();
    void perform();
}
```

Now, let us create two classes that both implement the same interface. As mentioned previously, my examples only show the code that demonstrates the concept. You will add whatever code is necessary for the class to do its work:

```
public class Version01 implements SharedInterface{
    @Override
    public String whatAmI() {
        return "Version 01";


    }
    @Override
    public void perform() {
        System.out.printf("Running perform in Version 01");
    }
}

public class Version02 implements SharedInterface {
    @Override
    public String whatAmI() {
        return "Version 02";
    }
    @Override
    public void perform() {
        System.out.printf("Running perform in Version 02");
    }
}
```

Now, we can look at the Factory class itself:

```
public class Factory {
    public static SharedInterface getInstance(
                                    String designator) {
```

Based on the string we pass to getInstance, we instantiate the appropriate object. Take note of default in the switch. It will return a null reference, a reference to nothing. You should test for this and take the appropriate action in the case where an invalid string is being used as designator:

```
        return switch (designator) {
            case "version01" -> new Version01();
            case "version02" -> new Version02();
            default -> null;
        };
    }
}
```

Now, let us look at code that will use the Factory pattern to instantiate the classes:

```
public class FactoryExample {
```

The objects we create will implement this interface:

```
    private SharedInterface version;
    public void perform(String versionName) {
```

Here, we pass the string that the Factory pattern will use to determine which class to instantiate:

```
        version = Factory.getInstance(versionName);
        System.out.printf(
            "Version: %s%n",version.whatAmI());
        version.perform();
    }
    public static void main(String[] args) {
        new FactoryExample().perform("version02");
    }
}
```

Using the factory pattern can simplify the creation of families of classes that all share the same interface.

Adapter

Imagine that you are working with existing code that uses a specific class with its own unique interface. One day, you come across another class that performs a similar task that meets the needs of a different client or is superior to what you have been using. The problem with the new class is that it does not have the same interface. Do you rewrite your code so that it will call upon the methods in the new class? You could, but the moment you start altering the existing code, there is the potential for unforeseen problems. The solution is to wrap the new class in an adapter class. The adapter presents the interface your code is already familiar with but then calls upon the appropriate method in the new class. This is where the adapter, a **structural** pattern, comes in.

Let us begin with a very simple application that calculates the fuel consumption of a vehicle and returns the result as miles per gallon. We begin with an interface for a class that will perform the calculation followed by its implementation:

```
public interface USFuelConsumption {
    String calculateUS(double distance, double volume);
}


public class USCar implements USFuelConsumption {
    @Override
    public String calculateUS(double distance,
                                    double volume) {
        return "MPG = " + distance/volume;
    }
}
```

Here is the code that will be used in this class:

```
public class AdapterExample {
    private USFuelConsumption consumption;
    public AdapterExample() {
        consumption = new USCar();
    }
    public void perform() {
        System.out.printf(
                "%s%n",consumption.calculateUS(350.0, 12.0));
    }
    public static void main(String[] args) {
        new AdapterExample().perform();
```

```
    }
}
```

The output of this program is as follows:

MPG = 29.166666666666668

Now imagine that a new client wants to use your system but needs the calculation done using metric measurement. For cars, this is described as liters per 100 kilometers. We have an interface and a class that will do this:

```java
public interface MetricFuelConsumptions {
    String calculateMetric(double distance, double volume);
}


public class MetricCar implements MetricFuelConsumptions {
    @Override
    public String calculateMetric(double distance,
                                        double volume) {
        return "l/100km = " + volume/distance * 100;
    }
}
```

To be able to use this new class, we need an adapter, which will implement the same interface but in the method call, will use the metric calculation class:

```java
public class UstoMetricAdapter implements USFuelConsumption {
    private final MetricCar metric;
    public UstoMetricAdapter() {
        metric = new MetricCar();
    }
```

Here is the method we are adapting. Rather than doing the calculation here, it will call upon the method of MetricCar:

```java
    @Override
    public String calculateUS(double distance,
                                    double volume) {
        return metric.calculateMetric(distance, volume);
    }
}
```

Now, let us see how it will be used:

```
public class AdapterExample {
    private USFuelConsumption consumption;
    public AdapterExample() {
```

Here is the only line we need to change. As the adapter shares the same interface, it can be used in place of USCar:

```
        consumption = new UstoMetricAdapter();
    }
    public void perform() {
        System.out.printf("%s%n",
            consumption.calculateUS(350.0, 44.0));
    }
    public static void main(String[] args) {
        new AdapterExample().perform();
    }
}
```

The program output is now as follows:

```
1/100km = 12.571428571428573
```

This is a trivial example, but it is an example of code reuse. The adapter allows you to reuse code expecting a specific interface with new code. This new code has a different interface, and the adapter resolves this by presenting the original interface to your code.

Observer

For this pattern, we are interested in when the state of an object changes. When a state change happens, a method in another class is called to carry out some tasks. These tasks could be to validate the state change, write the change to a database, or they could update the display. Java makes it easy to use this pattern by providing the PropertyChangeListener interface and the PropertyChangeSupport class. This is an example of a **behavioral** pattern.

We begin with a class that must notify other classes should the state of any or all of its fields change:

```
public class TheProperty {
```

This is the field in this class that we are planning to listen for changes to its state:

```
private String observedValue = "unicorn";
private final PropertyChangeSupport support;
```

The constructor is instantiating an instance of the PropertyChangeSupport class. This object will allow us to add to or remove from a list of all the classes that implement the PropertyChangeListener listener for this class. It supports firing an event when a field changes:

```
public TheProperty() {
    support = new PropertyChangeSupport(this);
}
```

This method lets us add to the list of listeners:

```
Public void addPropertyChangeListener(
            PropertyChangeListener listener) {
    support.addPropertyChangeListener(listener);
}
```

This method allows the removal of a listener:

```
Public void removePropertyChangeListener(
            PropertyChangeListener listener) {
    support.removePropertyChangeListener(listener);
}
```

This is the set method for the observedValue variable. When this method is called, firePropertyMethod will call propertyChange in every class that is in the list of listeners:

```
public void setObservedValue(String value) {
    System.out.printf(
            "TP: observedValue has changed.%n");
    support.firePropertyChange(
            "observedValue", this.observedValue, value);
    observedValue = value;
}
```

observedValue is a private field, as it should be. We need methods such as the previous set method and this get method to read the value in the field:

```
public String getObservedValue() {
```

```
        return observedValue;
    }
}
```

Now, we need a `listener` class. There can be just one or there can be many:

```
public class TheListener implements PropertyChangeListener{
```

This is the field that we wish to change if the field in `TheProperty` changes. While this is the usual way to use this pattern, the method that will be called when a change occurs is free to do whatever it wants and not just update a field in the listener:

```
    private String updatedValue;
```

Here is the method that is called in every listener to `TheProperty`. While it is assigning the new value from `TheProperty` to its own `updatedValue`, you could do anything in this method, even write to a database as an example. Notice that the `PropertyChangeEvent` object has access to a name you gave to the property, usually the field name, along with the previous and new values. The name can be used to decide on different actions depending on which field was changed:

```
    @Override
    public void propertyChange(PropertyChangeEvent evt) {
        System.out.printf("TL: The state has changed.%n");
        System.out.printf("TL: Observed field:%s%n",
                        evt.getPropertyName());
        System.out.printf('TL: Previous value: %s%n',
                        evt.getOldValue());
        System.out.printf("TL: New value: %s%n",
                        evt.getNewValue());
        setUpdatedValue((String) evt.getNewValue());
    }
```

This class also has a set and get method for its private `updatedValue` field:

```
    public String getUpdatedValue() {
        return updatedValue;
    }
    public void setUpdatedValue(String updatedValue) {
        this.updatedValue = updatedValue;
    }
}
```

Now we can test that if a change is made to the `TheProperty` object's field, then `TheListener` is notified:

```java
public class PropertyListenerExample {
    public void perform() {
```

We need at least one observable and any number of observers. They can be local variables like this or class fields:

```java
        var observable = new TheProperty();
        var observer = new TheListener();
```

A listener is added to the observed object:

```java
        observable.addPropertyChangeListener(observer);
```

Now, we update the observed field in the observable object. This will also result in the field in the observer object being updated:

```java
        observable.setObservedValue("moose");
        System.out.printf(
                "PLE: New value in observer is %s%n",
                observable.getObservedVable());
    }

    public static void main(String[] args) {
        new PropertyListenerExample().perform();
    }
}
```

In this section, we looked at four of the many patterns. Patterns themselves are broken down into categories. Singleton and factory are creational patterns. Adapter is a structural pattern. Observer is a behavioral pattern. All the patterns can be applied to any language you use.

Summary

In this chapter, we just touched upon SOLID software design principles and the singleton, factory, adapter, and observer design patterns. There are dozens of other principles and patterns. Design principles should guide your everyday coding, and design patterns offer solutions to design problems. Both are applicable to any language.

Next up, we will be looking at documenting and testing the code that you write.

Further reading

- *Design Principles in Java*: `https://www.javatpoint.com/design-principles-in-java`

- *Object Oriented Design Principles in Java*: `https://stackabuse.com/object-oriented-design-principles-in-java/`

- *What's a Software Design Pattern? (+7 Most Popular Patterns)* `https://www.netsolutions.com/insights/software-design-pattern/`

- *Design Patterns: Elements of Reusable Object-Oriented Software*, by Erich Gamma, John Vlissides, Ralph Johnson, and Richard Helm, 0-201-63361-2, available in libraries or from online booksellers

11

Documentation and Logging

In this chapter, we will look at two aspects of software development that do not directly influence the operation of the code. The first is documentation, more commonly called comments. The second is logging, a tool used to record events during the run of a program for the purpose of monitoring what the program is doing. We will begin with documenting code inline. You likely noticed that none of the code shown so far in this book has any comments. This has been done on purpose, as each chapter describes what the code is doing. If you look at the book's code in the GitHub repository, you will find comments in every file.

You have probably seen a message telling you to look at the log file when something goes wrong in a program. Where do these log files come from? We will examine how we can either display messages on the console or write to a file that certain events occurred or exceptions were thrown.

Here is the rundown for this chapter:

- Creating documentation
- Using logging

By the end of this chapter, you will understand the various ways in which comments are added to source code. You will also learn to use logging to record events in your code.

We will begin with documentation, but before that, let's have a quick look at the prerequisites for this chapter.

Technical requirements

Here are the tools required to run the examples in this chapter:

- Java 17
- A text editor
- Maven 3.8.6 or a newer version installed

The sample code for this chapter is available at `https://github.com/PacktPublishing/ Transitioning-to-Java/tree/chapter11`.

Creating documentation

Having been a computer science instructor for 31 years, I can tell you that the one task most students put off for as long as possible is documenting their code. I have learned of companies that forbid their developers to comment in their code. These companies believe that code should be self-documenting. If you cannot understand the purpose of the code from how it is written, then it has been written poorly. This is a big mistake. Students doing internships at such companies report spending an inordinate amount of time trying to understand the company's code base.

Documenting or commenting in code is not about explaining or apologizing for writing bad code. Code carries out tasks and the task should be obvious from the code itself. What is never obvious is why the code has been constructed in a certain way and how it may fit in with the rest of the program. One question to ask yourself is whether the programmer who takes over this code base when you get promoted or you move to another company understands what you coded and why.

Now let us look at how we add comments to our code and the unique commenting technique available in Java, called Javadocs.

Comments

In Java, there are three ways to indicate a comment in your code. The first is the original C style comment designation, which uses an opening forward slash, then an asterisk, a closing asterisk, then a closing forward slash as a set of characters:

```
/* This is a single line comment */
/* This is a multiple line
   Java comment */
```

Finally, there is the inline comment form:

```
System.out.println("Moose "/* + x */);
```

Here, you can use the `/* ... */` characters to comment out a section of code.

One important proviso is that you cannot nest these comments inside each other as shown:

```
/* Comment 1 /* Comment 2 */ end of comment 1 */
```

The Java compiler will see the first `/*` notation as ending at the first `*/` notation; leaving the end of the comment will likely be a syntax error.

The second set of characters you can use for comments is the double forward slash. These are single-line comments that end when the line ends. They can be placed anywhere and everything that follows becomes a comment:

```
// Meaningless example
int x = 4;
int z = 6;
int y = x * z; // Initializing y with x times z
```

These are also useful for commenting lines of code. Whenever I am making a change to an existing line of code, I first comment out the line I am replacing before I write the new line. I rarely delete code until I am certain that the new code is working.

There is one more way we can add comments to code and that is by creating Javadocs.

Javadocs

Javadocs are HTML pages created by the Javadoc tool, which is included in the Java installation. It examines every Java file and constructs an HTML page for each public class. These pages include all public fields and methods. While we will only look at this default behavior, you can adjust it. Although private elements are ignored by Javadocs, it is considered best practice to comment everything as if it were public.

Here is a sample program to which we will apply the Javadocs tool. The comments continue the discussion on Javadocs, so please do not skim over this but rather read it like every page in this chapter.

```
package com.kenfogel.javadocsexample;

/**
 * This is an example of using Javadocs in your source
 * code. It is only used for commenting a class, a method,
 * or a public field. It cannot be used inline. It begins
 * with a forward slash followed by two asterisks.
 *
 * You can inline certain HTML designations as follows:
 * <p>
 * <b>bold</b></p>
 * <p>
 * paragraph</p>
```

```
 * <p>
 * <i>italic</i></p>
 *
 * You can even create a list:
 * <ul>
 * <li>First item in the list
 * <li>Second item in the list
 * </ul>
 *
 * There is a lot of discussion on whether the
 * {@literal @version} tag should be used or whether the
   version should come from a repo such as Git.
 *
 * @version 1.0
 * @author Ken Fogel
 */
public class JavaDocsExample {
    /**
     * This is a private field so this comment will
     * not appear in the HTML file.
     */
    private final String name;
    /**
     * This is a public field, so this comment, written
     * above the field, will appear using /** to start
     * the comment.
     */
    public final int number;

    /**
     * In the method's Javadocs, list all the method
     * parameters with the {@literal @param} tag. That this
     * is also a constructor and is recognized as the
     * method has the same name as the class and does not
     * have a return type.
```

```
 *
 * @param name: The user's name
 * @param number: The answer to the ultimate question
 */
public JavaDocsExample(String name, int number) {
    this.name = name;
    this.number = number;
}

/**
 * While you can and should comment private methods as
 * Javadocs, they will not appear on the HTML page.
 * Only public methods appear in the Javadocs.
 *
 * @param day The day of the week that will be
 * displayed
 * @return The string to display
 */
private String constructMessage(String day) {
    return name + " " + number + " " + day;
}

/**
 * Here is a public method whose Javadoc block will
 * appear in the HTML.
 *
 * @param day The day of the week
 */
public void displayTheMessage(String day) {
    System.out.printf('%s%n", constructMessage(day));
}

/**
 * Here is the method where the program will begin:
 *
```

```
     * @param args values passed at the command line
     */
    public static void main(String[] args) {
        new JavaDocsExample(
                "Ken", 42).displayTheMessage("Wednesday");
    }
}
```

To run the javadoc tool, use the following command line with switches:

```
javadoc -d docs
```

The -d switch is the location to which the HTML files will be written. In this case, it is assumed that there is a folder named docs in whatever folder you are currently in. The folder must exist, as javadoc will not create it. If the folder does not exist, then the HTML files will be written into the current folder:

```
javadoc -d docs -sourcepath C:\dev\PacktJava\Transitioning-to-
java\JavadocsExample\src\main\java
```

The -sourcepath switch is the path to the folder that contains either Java files or packages. As this is a Maven-based project, the packages and source files are always found in \src\main\java in whatever folder the Maven project is in:

```
javadoc -d docs -sourcepath C:\dev\PacktJava\Transitioning-to-
Java\JavaDocsExample\src\main\java -subpackages com:org
```

The last switch, -subpackages, is a colon-separated list of packages in the project. javadoc will recursively go through every folder and subfolder, starting with the names in the list, to find Java files to be processed. I created a second package that began with org. -subpackages are searched recursively and all public or package classes found in any folder starting with the listed names will be documented.

When the javadoc tool is run on the project, it will create HTML web pages. What follows is the Javadocs web page created for the JavaDocsExample class. It can be quite long. Take note that only public methods appear. Private methods, though commented like a public method, do not appear in the HTML output. Here is what the Javadocs will look like.

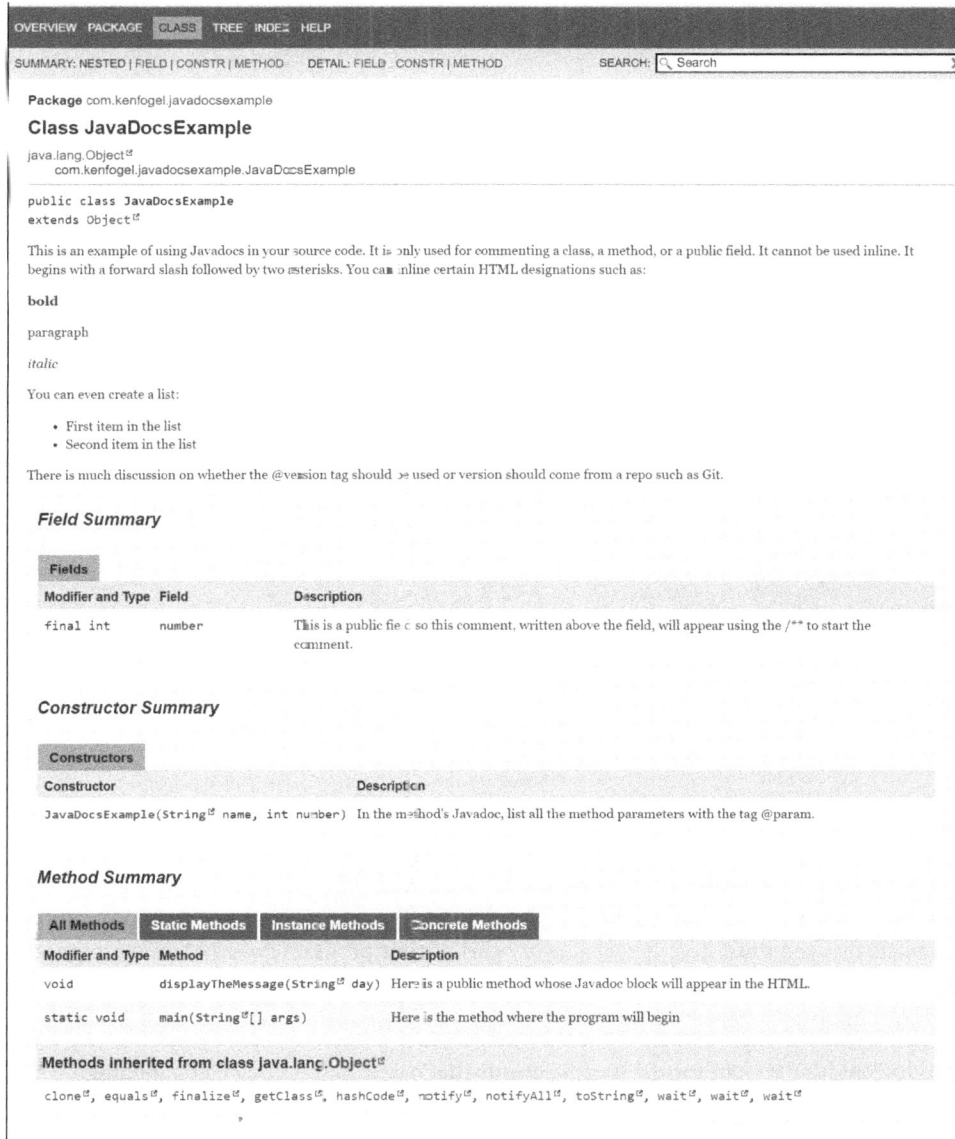

Figure 11.1 – The first half of the generated Javadocs

Field Details

number

```
public final int number
```

This is a public field so this comment, written above the field, will appear using the /** to start the comment.

Constructor Details

JavaDocsExample

```
public JavaDocsExample(String name,
                       int number)
```

In the method's Javadoc, list all the method parameters with the tag @param. That this is also a constructor is recognized as the method has the same name as the class and does not have a return type.

Parameters:

name - user's name

number - answer to the ultimate question

Method Details

displayTheMessage

```
public void displayTheMessage(String day)
```

Here is a public method whose Javadoc block will appear in the HTML.

Parameters:

day - day of the week

main

```
public static void main(String[] args)
```

Here is the method where the program will begin

Parameters:

args - values passed at the command line

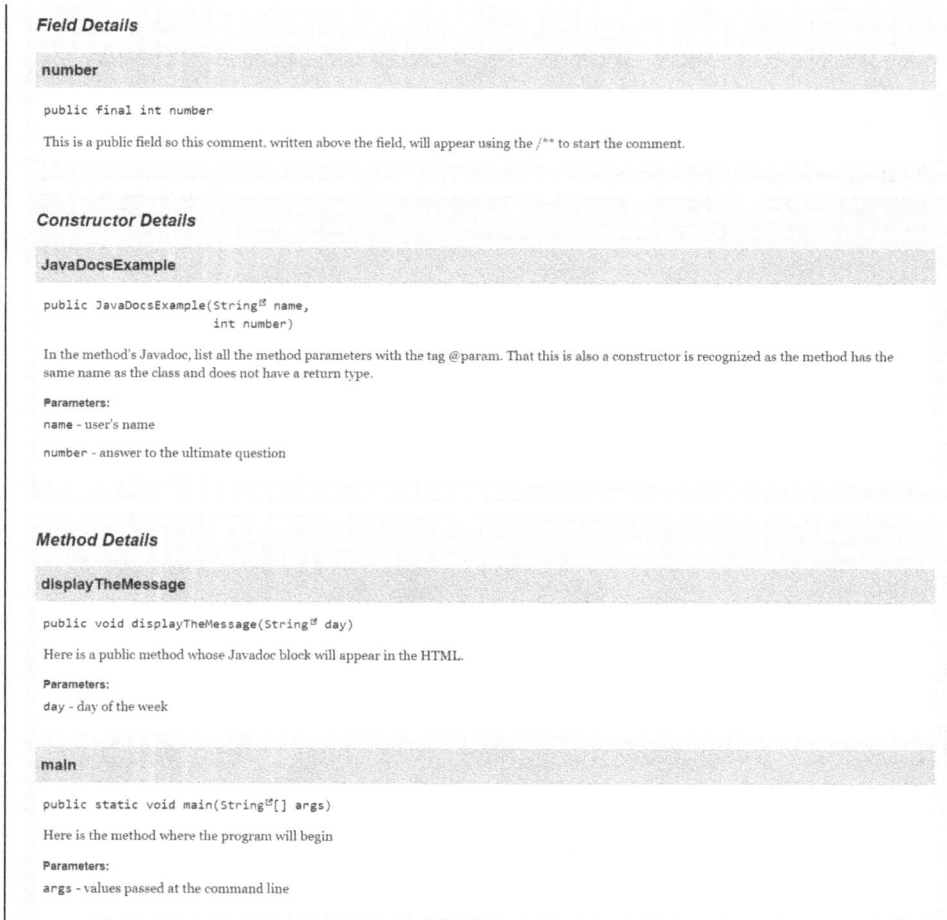

Figure 11.2 – The second half of the generated Javadocs

The entire Java library is described in Javadocs and is searchable in your browser. See *Further reading* for the URL to these docs. The best practice for the code that you write is to write Javadocs comments. This also means that you must describe what every part of your program does and, more importantly, why it does what it should. Use the /* . . . */ and // notations to include additional comments in methods or temporarily remove code.

Now, let us look at how we can record specific events that occur in our code using logging.

Using logging

In your own code, you may want to display messages in the console while the program runs. These messages may be to inform you that an exception has been caught or record any other event that happens during the program's execution. While you can write to the console using System.out.print, println, or my favorite, printf, do not. If the application is console-based, then these statements will appear with the console user interface. For GUI or web applications, the console may or may not be visible. Once the program goes into production, the end user may be confused or overwhelmed by the messages you display in the console.

The solution is logging. This allows you to write log messages to the console, a file, or a database, or even send them to yourself in an email. We will only look at the console or a file. Java has a logging framework, found in java.util.logging. We will also look at one of the external logging frameworks from the Apache Foundation, called **Log4j2**.

java.util.logging

There are two parts to a logging framework. There is the framework of Java classes and the configuration file. For JUL, the common name for java.util.logging, there is a configuration file named logging.properties in the conf folder of the Java installation. We will see how to use a custom config file rather than use the config shared by all applications. The default location for the JUL config is in the Java conf folder. We can place our JUL custom properties file anywhere on our system, as we must provide the path to the file when we instantiate the logger.

Here is a simple program that uses the logger:

```
public class JULDemoDefaultConfig {
```

We instantiate Logger using the Factory Software pattern as implemented in the Logger class. We pass the name of this class so that it can appear in the logger output and we can support different Logger items for different classes:

```
    private static final Logger LOG =
    Logger.getLogger(JULDemoDefaultConfig.class.getName());
```

Log messages must be associated with a level, which is the first parameter when we use the log method. There are six levels, and all have an optional third parameter of an Exception object. Typically, the Level.INFO parameter is used for recording information that you wish to record about what the program is doing or who is doing it. Level.SEVERE is used for recording exceptions. The FINEST, FINER, and FINE parameters are used while debugging an application. You can decide the minimum level in the config file. During development, you will use ALL, while once put into production you will, raise the level to INFO. This means that you do not need to delete or comment out log messages below INFO.

In this method, we just create log messages:

```
    public void perform() {
        LOG.log(Level.FINEST,
```

```
        "JUL default-Using LOG.log at Level.FINEST");
    LOG.log(Level.FINER,
        "JUL default-Using LOG.log at Level.FINER");
    LOG.log(Level.FINE,
        "JUL default-Using LOG.log at Level.FINE");
    LOG.log(Level.INFO,
        "JUL default-Using LOG.log at Level.INFO");
    LOG.log(Level.WARNING,
        "JUL default-Using LOG.log at Level.WARNING");
```

For the SEVERE level, I have forced an exception in the try block and when it is caught, I log it by including the Exception object.

```
        try {
```

You can add a custom message to an exception by passing a String object to the constructor:

```
        throw new Exception(
                "JUL default config exception");
        } catch (Exception ex) {
            LOG.log(Level.SEVERE,
        "JUL default-Using LOG.log at Level.SEVERE", ex);
        }
    }
    public static void main(String[] args) {
        new JULDemoDefaultConfig().perform();
    }
}
```

If you have a custom config file, you must explicitly load the file; otherwise, the default config, logging.properties, in the Java conf folder will be used. It is not a good idea to change the default configuration, as it will affect every program that you run that uses JUL.

To load a custom config file, you need to find this:

```
private static final Logger LOG =
    Logger.getLogger(JULDemoDefaultConfig.class.getName());
```

Replace it with the following:

```
private static final Logger LOG;
static {
```

When your code is packaged into a JAR file, the location of resource files that existed in src/main/resources is the root of the project. The retrieveURLOfJarResource method knows this, so it can load the config file placed in this folder. This is in a static initialization block, which will ensure that this Logger class will only be instantiated once should there be more than one instance of this class:

```
try (InputStream is =
        retrieveURLOfJarResource("logging.properties").
        openStream()) {
    LogManager.getLogManager().readConfiguration(is);
} catch (Exception e) {
    Logger.getAnonymousLogger().severe(
            "Unable to load config\nProgram is exiting");
    System.exit(1);
 }
 LOG = Logger.getLogger(
    JULDemoCustomConfig.class.getName());
}
```

The default logging.properties file is very well commented. Here are the contents of the file with the comments removed. I encourage you to examine the version of this file on your machine.

Here's the display output to the console:

```
handlers= java.util.logging.ConsoleHandler
```

Unless it's overridden, this will only show logs of this level or greater:

```
.level= INFO
```

If you are writing logs to a file, then the pattern property is the folder and filename. In this case, %h means to write the file to your home directory. This is the best practice for JUL. If you prefer to store the log files in a specific folder name, then it must already exist:

```
java.util.logging.FileHandler.pattern = %h/java%u.log
java.util.logging.FileHandler.limit = 50000
```

Every time the program runs, it overwrites the previous log file, as only one log file is permitted:

```
java.util.logging.FileHandler.count = 1
```

Logging is thread-safe. This tells us that up to 100 concurrent log file locks can be used. If you get IOException errors when writing to a log, you may solve this issue by increasing the number of locks:

```
java.util.logging.FileHandler.maxLocks = 100
```

Write to the log file in the XML format:

```
java.util.logging.FileHandler.formatter =
                    java.util.logging.XMLFormatter
```

Overriding a handler's level supersedes the global level:

```
java.util.logging.ConsoleHandler.level = INFO
```

This is the format of the log when displayed on the screen. You can configure SimpleFormatter, and this is explained in the comments for the default config file:

```
java.util.logging.ConsoleHandler.formatter =
                    java.util.logging.SimpleFormatter
```

The custom properties file has the following changes:

- The FileHandler class has been added so that logs will be written to a file and the console:

```
handlers= java.util.logging.FileHandler,
                java.util.logging.ConsoleHandler
```

- Both handlers will now display log messages of every level:

```
java.util.logging.ConsoleHandler.level = ALL
java.util.logging.FileHandler.level = ALL
```

We use %h to indicate that we want the logs written to our home directory. If you wish to write them to a specific folder, then the folder must already exist. If the folder does not exist, then the file will not be created:

```
java.util.logging.FileHandler.pattern =
                        %h/loggingdemo-JUL_%g.log
```

There can be three log files, one for each run of the program. After writing to the third log file, should there need to be another log file, then it wraps around and overwrites the existing files in the order they were created:

```
java.util.logging.FileHandler.count = 3
```

The Java logger is always available and does not have any dependencies that you must add to the Maven POM file. Logging can have an impact on the performance of your code. For this reason, there are alternatives to JUL that execute in less time or provide features not present in JUL. Let's look at one of the most widely used external loggers, Log4j2.

Log4j2

Log4j2 works very much like JUL. Before we can use it, we need to add new dependencies to our POM file:

```
<dependency>
    <groupId>org.apache.logging.log4j</groupId>
    <artifactId>log4j-api</artifactId>
    <version>2.19.0</version>
</dependency>
<dependency>
    <groupId>org.apache.logging.log4j</groupId>
    <artifactId>log4j-core</artifactId>
    <version>2.19.0</version>
</dependency>
```

In any file in which you plan to use Log4j2, you begin with this class field:

```
private static final Logger LOG =
                LogManager.getLogger(Log4jDemo.class);
```

Now to be able to log, you just need the following. Take note that the levels are now methods of the LOG object. An optional second parameter can take an Exception reference for all levels, as shown in Level 5:

```
public void perform() {
    LOG.trace("log4j2-Level 1: I am a trace");
    LOG.debug("log4j2-Level 2: I am a debug");
    LOG.info("log4j2-Level 3: I am an info");
    LOG.warn("log4j2-Level 4: I am a warning");
    try {
        throw new Exception("log4j2 exception");
    } catch (Exception ex) {
        LOG.error("log4j2-Level 5: I am an error", ex);
    }
```

When using Log4j2, you should create a config file, as its default behavior is limited. In the absence of this file, the logger will do nothing. Like JUL's config, the `log4j2.xml` config file is expected to be found in `src/main/resources`.

Rather than review this file, I ask you to clone the repo from GitHub for this chapter and look at the `log4j2.xml` file in the `LoggingExample` project. Its comments explain what can be configured. One improvement over JUL is that should you wish to store the logs in an arbitrary folder, Log4j2's file handler will create the folder.

I add a logger to almost every file I write. This allows me to write logs as needed. Declaring a `Logger` object that you do not use will have no effect on the performance of your program.

Summary

In this chapter, we covered two important tasks that every programmer should include in their code, regardless of the language used. The first was documentation. Comments and Javadocs can be critical in the maintenance of existing code or in adding new features. You may think you will never forget why you coded in a certain way, but 6 months from now, that memory may not be as accurate as it needs to be.

During the development of software, and once it goes into production, having the program write what it is doing to the console, or more commonly, to a file, can go a long way in tracking down bugs. Auditing software that is subject to regulations is another task logging can carry out. Never use `System.out.print` or its cousins to display information about the operation of a program – use a logger. Either the Java logger or an external logger such as Log4j2 should be, must be deployed in your code.

Documenting your code is mandatory. Using logging to record events in a program is mandatory. Remember that programming is an engineering discipline and not an art form. Engineering requires the type of documentation described here and requires the use of logging to monitor a program's performance.

Coming up, we will look at how to work with floating-point numbers when absolute accuracy is required. Testing our code to ensure it performs as designed is also covered in the next chapter.

Further reading

- *How to Write Doc Comments for the Javadoc Tool*: `https://www.oracle.com/technical-resources/articles/java/Javadoc-tool.html#styleguide`

- *The Java 17 Javadocs*: `https://docs.oracle.com/en/java/javase/17/docs/api/index.html`

- *Java Logging Tools and Frameworks*: `http://www.java-logging.com/`

12
BigDecimal and Unit Testing

We begin this chapter by addressing the problem with floating point representation that is found in most languages. The problem revolves around the inability to represent every decimal fraction as a binary fraction, as pointed out in *Chapter 4, Language Fundamentals – Data Types and Variables*. In most situations, it can be accurate enough. But what happens if you must guarantee accuracy and precision? You must abandon floating point primitives and use the `BigDecimal` class.

How do you know that the code you have just written works? The compiler can spot syntax errors. An error-free compilation only tells you that the compiler is happy. But does it work? How does your code handle invalid input, lost connections to a database, or edge cases? Always be aware that for most projects you work on, the most unreliable component of the systems you code for is the end users. You cannot fix them, but you need to design and implement your code to handle the unexpected. Unit testing is one technique for validating your code while you write it.

Unit testing is not the same as **quality assurance (QA)**. This is a process carried out to ensure that the program is meeting the specifications it was coded for. QA is about programs that run. Unit testing is about the performance of individual methods. Sometimes objects and methods that work together must be tested, which is called integration testing, but the testing techniques are similar. This testing is the responsibility of the programmer.

In this chapter, we will look at how we write unit tests using the JUnit 5 framework. What is significant about this framework is that you can test any method in your code without the need for a main method. We will cover the following topics:

- Using `BigDecimal`
- What is JUnit 5?
- Testing with JUnit 5
- Performing parametrized testing

In the advanced Java courses that I have taught, unit testing was mandatory. If a student could not demonstrate that the code they wrote passed unit tests, then I could not be bothered to even look at the code. A submission was expected to run its tests before the program itself was executed. If there were no tests, then it was an automatic failure. I may have been harsh, but the result was that I had confidence that what they coded could work.

Technical requirements

Here are the tools required to run the examples in this chapter:

- Java 17
- A text editor
- Maven 3.8.6 or a newer version installed

The sample code for this chapter is available at https://github.com/PacktPublishing/Transitioning-to-Java/tree/chapter12.

Using BigDecimal

The BigDecimal class, which is a member of the java.math library, is a fixed-precision representation of floating point numbers. This means that values represented as BigDecimal do not suffer from the problem of approximation that can and does occur when calculations are carried out by the hardware **floating point unit** (**FPU**) of most CPUs.

The BigDecimal class shares an important characteristic with a string. They are both immutable. This means that when a value becomes a BigDecimal object, it cannot be changed. Any operation on a BigDecimal object returns a new BigDecimal object.

Let us look at an application that can calculate loan payments for money borrowed. The formula for this calculation is as follows:

$$PMT = PV \times \frac{rate}{1 - (1 + rate)^{-n}}$$

Here:

- $rate$ = the interest rate per period
- n = the number of periods
- PV = present value (amount of loan)
- PMT = payment (the monthly payment)

If we use doubles for all the values, the Java bean data object will look as follows:

```
public class FinancialData {
    private double amountBorrowed;
```

```
    private double annualRate;
    private double term;
    private double monthlyPayment;
    public FinancialData(double amountBorrowed,
            double annualRate,
            double term) {
        this.amountBorrowed = amountBorrowed;
        this.annualRate = annualRate;
        this.term = term;
        this.monthlyPayment = 0.0;
    }
    public FinancialData() {
        this(0.0, 0.0, 0.0);
    }
    public double getAnnualRate() {
        return annualRate;
    }
    public void setAnnualRate(double annualRate) {
        this.annualRate = annualRate;
    }
// There are setters and getters for
// the other three fields.
```

The Object superclass has a method named toString that will return the address at which the object is stored as a string. We override it to display the values in all the fields. This can be quite useful in debugging, so I advise you to always have a toString method in any data class:

```
    @Override
    public String toString() {
            return "FinancialData{" + "amountBorrowed=" +
            amountBorrowed + ", annualRate=" +
            annualRate + ", term=" + term +
            ", monthlyPayment=" + monthlyPayment + '}';
    }
```

The formula to calculate the result is as follows. It has been broken up to reflect each part of the final calculation, although it could be written as a single line as done in the CompoundInterest examples. The comments describe each part of the formula:

```java
public class Calculation {
    public void loanCalculator(FinancialData data) {
        // Convert APR to monthly rate because payments are
        // monthly
        var monthlyRate = data.getAnnualRate() / 12.0;
        // (1+rate)
        var temp = 1.0 + monthlyRate;
        // (1+rate)^term
        temp = Math.pow(temp, -data.getTerm());
        // 1 - (1+rate)^-term
        temp = 1 - temp;
        // rate / (1 - (1+rate)^-term)
        temp = monthlyRate / temp;
        // pv * (rate / 1 - (1+rate)^-term)
        temp = data.getAmountBorrowed() * temp;
        data.setMonthlyPayment(Math.abs(temp));
    }
}
```

If we borrowed $5,000 at 5% annual interest for 60 months, the answer will be $94.35616822005495. So, the result should be $94.36. The issue here is that all calculations are being done to 14 decimal places when they should only be done with values that have two decimal places except for the monthly interest rate. Interest rates may have more than two decimal places. Dividing the annual interest rate by 12, for 12 payments a year, results in a value where the first two decimal places are 0. In most cases, the result will be accurate, but not always. This is a serious issue if you are writing what is referred to as an accounting problem. The solution is to use BigDecimal. Here is the data object:

```java
public class FinancialData {
    private BigDecimal amountBorrowed;
    private BigDecimal annualRate;
    private BigDecimal term;
    private BigDecimal monthlyPayment;
    public FinancialData(BigDecimal amountBorrowed,
            BigDecimal annualRate,
            BigDecimal term) {
        this.amountBorrowed = amountBorrowed;
        this.annualRate = annualRate;
        this.term = term; .
```

There are convenience objects in `BigDecimal`, one of which is `BigDecimal.ZERO`, which returns a `BigDecimal` object initialized to 0:

```
        this.monthlyPayment = BigDecimal.ZERO;
    }
```

The default constructor is using this non-default constructor and passes it three `BigDecimal` objects initialized to 0:

```
    public FinancialData() {
        this(BigDecimal.ZERO, BigDecimal.ZERO,
            BigDecimal.ZERO);
    }
    public BigDecimal getAnnualRate() {
        return annualRate;
    }
    public void setAnnualRate(BigDecimal annualRate) {
        this.annualRate = annualRate;
    }
// There are setters and getters for the other three
// fields along with a toString method.
```

The `Calculation` class using `BigDecimal` now looks as follows:

```
public class Calculation {
    public void loanCalculation(FinancialData data)
                throws ArithmeticException {

        var monthlyRate = data.getAnnualRate().
                    divide(new BigDecimal("12"),
                    MathContext.DECIMAL64);
        // (1+rate)
        var temp = BigDecimal.ONE.add(monthlyRate);
        // (1+rate)^term
        temp = temp.pow(data.getTerm().intValueExact());
        // BigDecimal pow does not support negative
        // exponents so divide 1 by the result
```

Division is an operation that could result in an infinitely repeating sequence. If this is detected, then an exception is thrown. To prevent this exception, we limit the number of decimal places with `MathContext.DECIMAL64`. This will limit the number to 16 decimal places:

```
temp = BigDecimal.ONE.divide(
        temp, MathContext.DECIMAL64);
// 1 - (1+rate)^-term
temp = BigDecimal.ONE.subtract(temp);
// rate / (1 - (1+rate)^-term)
temp = monthlyRate.divide(
        temp, MathContext.DECIMAL64);
// pv * (rate / 1 - (1+rate)^-term)
temp = data.getAmountBorrowed().multiply(temp);
```

Here, we use `setScale` to restrict the output to two decimal places. We also define how rounding should occur. Many of you will have been taught that 1 to 4 round down and 5 to 9 round up. This is not how it is done in accounting. Banks use `HALF_EVEN`. For example, 27.555 will round to 27.56. If the value is 27.565, it will round to 27.56. If the value of the last requested decimal place is an even number and the value that follows is exactly 5, then it rounds down. If it is an odd number, it rounds up. Over time, you and the bank will break even. Without `HALF_EVEN`, you will probably lose money to the bank:

```
temp = temp.setScale(2, RoundingMode.HALF_EVEN);
```

Some financial calculations return a negative answer. This tells you which way the money flows, to you or to the bank. I am using the `BigDecimal` absolute method to eliminate the sign:

```
        data.setMonthlyPayment(temp.abs());
    }
}
```

The question now is how we can test this code to ensure it is giving us the right answer. We could write code in the `main` method to test it, as follows:

```
public static void main(String[] args) {
    var data = new FinancialData(
            new BigDecimal("5000.0"),
            new BigDecimal("0.05"),
            new BigDecimal("60.0"));
    new Calculation().loanCalculation(data);
```

As already noted in *Chapter 8, Arrays, Collections, Generics, Functions, and Streams*, you cannot use operators such as +, >, and ==, with objects. Instead, you use methods such as equals:

```
    if (data.getMonthlyPayment().equals(
            new BigDecimal("94.36"))) {
        System out.printf("Test passed%n");
    } else {
        System.out.printf("Test failed: %.2f %s%n",
                data.getMonthlyPayment(), "94.36");
    }
}
```

What if you want to test many values? The loanCalculation method shows that it could throw ArithmeticException. How can you test that this exception is thrown when appropriate? The answer is unit testing.

What is JUnit 5?

JUnit 5 is an open source library and not part of the Java Development Kit library. It is licensed using the Eclipse Public License v2.0. This simply means that you are free to use this library and distribute it with your work for either open source or commercial software without having to make any payments. So, what does it do?

This unit testing framework allows you to instantiate any class in your project and call upon any non-private method. These non-private methods, such as public and package, can be executed from within what is called a test class. These are classes that are instantiated by the JUnit framework. A test class contains methods that can instantiate any class in the project and call upon a method in the class.

Testing with JUnit 5

Test classes are not part of the usual source code folder, src/main/java. Instead, they are placed in src/test/java. They can and should be organized into packages. You can also have resources that are just used by the test classes, such as logging.properties or log4j2.xml. They will be placed in src/test/resources.

We will need to add new components to our Maven pom.xml file. The first is the dependency for JUnit 5. The first addition is the **bill of materials (BOM)**, in a section called dependencyManagement:

```
    <dependencyManagement>
        <dependencies>
            <dependency>
```

```
            <groupId>org.junit</groupId>
            <artifactId>junit-bom</artifactId>
            <version>5.9.1</version>
            <type>pom</type>
            <scope>import</scope>
        </dependency>
    </dependencies>
</dependencyManagement>
```

All dependencies and plugins in a Maven-based project require a version value. One way you can ensure that the version value for all dependencies of a given library is correct is to use, if available, a BOM. Now, it is no longer necessary to include version values for each library.

Next is the dependency specifically for JUnit 5. This dependency supports both single test methods and parameterized tests. Notice also that the scope of these dependencies is test, which means that they are not included in the final packaging of the code. The test classes are also not included in the final packaging of your code:

```
<dependency>
    <groupId>org.junit.jupiter</groupId>
    <artifactId>junit-jupiter</artifactId>
    <scope>test</scope>
</dependency>
```

The next change is to add the Maven surefire plugin. This plugin will run all unit tests. The results of the test will appear in the console, a text file, and an XML file. These files can be found in target/surefire-reports. This folder will be created for you when you run the tests. Existing test reports will be overwritten when the tests are rerun:

```
        <plugin>
            <groupId>org.apache.maven.plugins</groupId>
            <artifactId>
                maven-surefire-plugin
            </artifactId>
            <version>2.22.2</version>
        </plugin>
```

The example code does not have a main method as it represents a project in progress. It cannot be run but it can be unit tested. To just run the tests when you use mvn, set defaultGoal in the build section:

```
<defaultGoal>test</defaultGoal>
```

Let's create a basic unit test. The first thing you will need is to add the `test/java` and `test/resources` folders to the Maven project you are writing unit tests for. This is what the folder structure looks like for my example project. I have also added a package called `com.kenfogel.calculationtest` to `test/java`.

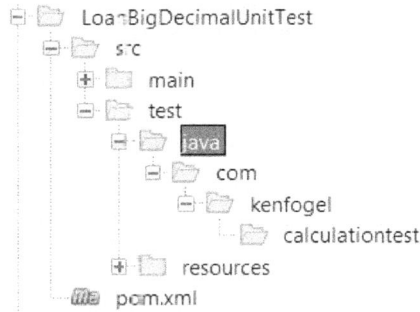

Figure 12.1 – Folder structure for unit tests

Now, let us look at our test class. The code examples in the book up to this point have not shown the required imports. They can be found in the code samples in the GitHub repo for the book. These examples will look at imports. Here is the `SimpleTest.java` class:

```
package com.kenfogel.calculationtest;
```

These two imports make the `Calculation` and `FinancialData` classes available to this class:

```
import
   com.kenfogel.loanbigdecimalunittest.business.Calculation;
import
   com.kenfogel.loanbigdecimalunittest.data.FinancialData;
```

Here are the imports for what we will use from JUnit 5:

```
import org.junit.jupiter.api.BeforeEach;
import org.junit.jupiter.api.Test;
```

This last `static` import allows us to use `assertEquals` without needing to show the whole package structure seen here:

```
import
   static org.junit.jupiter.api.Assertions.assertEquals;
```

Finally, we have the import for `BigDecimal`:

```
import java.math.BigDecimal;


public class SimpleTest {
    private Calculation calc;
    private FinancialData data;
```

The `@BeforeEach` annotation is used to define methods that must be run before each test method. It has a companion `@AfterEach` annotation. There are also `@BeforeAll` and `@AfterAll`, which are methods run once before all testing begins or after all testing ends. The best practice in testing is to always instantiate the objects you will use for testing for each test. Avoid reusing an object that was used in a previous test as it can result in unexpected errors in tests that depend upon it:

```
@BeforeEach
public void init() {
    calc = new Calculation();
    data = new FinancialData();
}
```

Here is the test annotated with `@Test`. It sets the three variables in `FinancialData` and calls upon the `Calculation` class to calculate the loan payment. It ends with `assertEquals` to compare the result with the known answer:

```
@Test
public void knownValueLoanCalculationTest () {
    data.setAmountBorrowed(new BigDecimal("5000"));
    data.setAnnualRate(new BigDecimal("0.05"));
    data.setTerm(new BigDecimal("60"));
    calc.loanCalculation(data);
    assertEquals(new BigDecimal("94.36"),
        data.getMonthlyPayment());
}
}
```

To run the test with the default goal set to `test`, you just need to run mvn at the command line in the root folder of the project. Here is the output of the test in the console:

```
[INFO] ------------------------------------------------------------
[INFO]  T E S T S
[INFO] ------------------------------------------------------------
[INFO] Running com.kenfogel.calculationtest.SimpleTest
[INFO] Tests run: 1, Failures: 0, Errors: 0, Skipped: 0, Time elapsed: 0.046 s -
in com.kenfogel.calculationtest.SimpleTest
[INFO]
[INFO] Results:
[INFO]
[INFO] Tests run: 1, Failures: 0, Errors: 0, Skipped: 0
[INFO]
[INFO] ------------------------------------------------------------
[INFO] BUILD SUCCESS
[INFO] ------------------------------------------------------------
[INFO] Total time:  1.497 s
[INFO] Finished at: 2022-12-21T17:10:09-05:00
[INFO] ------------------------------------------------------------
```

Figure 12.2 – The test result

You can also test that an expected exception is thrown. Here is the method rewritten but with `term` set to 0. This should result in `ArithmeticException` with a message of `Division by zero`. Asserting on the message is important as there are two possible reasons for `ArithmeticException`. The first is division by zero. The second occurs when a calculation using `BigDecimal` has an infinitely repeating sequence.

First, we need another `import` statement:

```
import static
      org.junit.jupiter.api.Assertions.assertThrowsExactly;
```

Now we can write the test:

```
@Test
public void knownValueLoanExceptionTest() {
    data.setAmountBorrowed(new BigDecimal("5000"));
    data.setAnnualRate(new BigDecimal("0.05"));
    data.setTerm(new BigDecimal("0"));
```

Here, we are calling the method we expect to throw an exception in `assertThrowsExactly`. This method begins with the name of the exception class we are expecting followed by a lambda expression to invoke the method we expect to throw `ArithmeticException`. The `assertThrowsExactly` method returns the exception object that was thrown, and we assign it to an `ArithmeticException` object. We can now use the `assertEquals` method to determine whether division by zero was the cause of this exception. If no exception is thrown or a different message is found, then the test will fail:

```
ArithmeticException ex =
      assertThrowsExactly(ArithmeticException.class,
```

```
              () -> {calc.loanCalculation(data);});
        assertEquals("Division by zero", ex.getMessage());
}
```

This concludes our look at basic unit testing where each test runs just once. Ideally, a unit test should be run with a range of values and not just one. This is what we will look at next.

Performing parameterized testing

This leaves one more type of testing to look at, a parameterized test. As you may have realized, if you want to run a test to determine whether the result is accurate for several values, then you will need one method per set of values. JUnit 5 simplifies this task by allowing you to create a list of values. Let's see how this works. Here is the new parameterized test class:

```
public class ParameterizedTests {
    private Calculation calc;
    private FinancialData data;
```

We will not instantiate the `FinancialData` object here as we did in the previous example. It will be created by a private helper method:

```
    @BeforeEach
    public void init() {
        calc = new Calculation();
    }
```

The first annotation declares that this will be a parameterized test. This means that this method will be run once for every row of data listed as part of @CsvSource:

```
    @ParameterizedTest
    @CsvSource({
        "5000, 0.05, 60, 94.36",
        "3000, 0.05, 24, 131.61",
        "20000, 0.05, 72, 322.10"
    })
```

The `ArgumentsAccessor` parameter will contain the current row of data to test. This method will be called for every row in @CsvSource:

```
    public void knownValueLoanCalculationTest_param (
                    ArgumentsAccessor args) {
```

Our data class wants `BigDecimal` values. To accomplish this, we have a private method called `buildBean` that receives an object of type `ArgumentsAccessor` and then converts that into a `FinancialData` object:

```
data = buildBean(args);
calc.loanCalculation(data);
```

Each row of CSV data has the answer as the last element. We are comparing the result stored in `monthlyPayment` with the last parameter:

```
assertEquals(new BigDecimal(args.getString(3)),
             data.getMonthlyPayment());
  }
```

Our helper method constructs a `FinancialData` object from the first three items in the `ArgumentsAccessor` object:

```
private FinancialData buildBean(ArgumentsAccessor args) {
   return
     new FinancialData(new BigDecimal(args.getString(0)),
                       new BigDecimal(args.getString(1)),
                       new BigDecimal(args.getString(2)));
   }
 }
```

There is still more to learn about unit testing. If a test depends on a specific object supplying a specific value to the test method but is not considered to be a point of failure, you can fake it. This is called mocking an object. You create the mock and what its method must return. See the *Further reading* section for a link to one of the widely used mocking libraries, called Mockito.

Summary

The first topic we covered in this chapter was the `BigDecimal` class. Floating point values as processed by modern FPUs have issues when moving from decimal to binary and back again. This can be critical in the field of accounting where every penny must balance. As a class, `BigDecimal` objects are not as easy to use as primitives, but it is the need for absolute accuracy that mandates their use.

As I stated at the start of this chapter, testing is a critical task that every programmer should be doing. You should be delivering code that works as expected in almost every situation it is used for. Unit tests do not prove that the program's logic is necessarily correct. This is usually tested by the QA team who are testing the execution of the program.

While writing this book, I came across a research paper that looked at unit testing. You can find the link in the *Further reading* section. It focused on Java and C# developers who used integrated development environments such as Visual Studio or IntelliJ. What the paper discovered is that less than half the developers in the study did any kind of software testing despite the ease with which it can be written in an IDE. Please do not be in the wrong half.

In this chapter, we looked at two separate concepts. The first was how we can perform calculations, such as in accounting, that must be accurate to a specific number of decimal places. We accomplish this by using the `BigDecimal` class to represent floating point numbers rather than `float` or `double`.

The second concept introduced software testing using unit tests. As a programmer, you need to be able to show that the public methods you write perform as expected. This is what unit testing is about. I wrote in the previous chapter that commenting and logging were mandatory. I add unit testing to the list of mandatory tasks a programmer is expected to perform.

Next, let us move on and look at the **user experience**, also referred to as **UX**. Up until now, sample code used a console UX akin to the output of a typewriter. In the coming chapter, we will look at the graphical user interface UX that Java has available for us.

Further reading

- JUnit 5: `https://junit.org/junit5/`

- *Developer Testing in The IDE – Patterns, Beliefs, And Behavior*: `https://repository.tudelft.nl/islandora/object/uuid:a63e79e0-e0e4-48cd-81ed-87f163810538/datastream/OBJ/download` (this is a PDF file and not a web page)

- Mokito – Tasty mocking framework for unit tests in Java: `https://site.mockito.org/`

Part 3:
GUI and Web Coding in Java

With the fundamentals out of the way, it is time to look at a Java application. In this part, we see how a business process used in the previous section can be used in programs that share the logic but that use a different GUI library from desktop application to web application.

This part contains the following chapters:

- *Chapter 13, Desktop Graphical User Interface Coding with Swing and JavaFX*
- *Chapter 14, Server-Side Coding with Jakarta*
- *Chapter 15, Jakarta Faces Application*

13

Desktop Graphical User Interface Coding with Swing and JavaFX

This chapter will introduce a simple but complete application for providing three common financial calculations. These are loan payments, the future value of money, and a savings goal. We will look at two versions of this application, one written using the Swing library and the second using the JavaFX library. The server-side coding for this application will be covered in *Chapter 15, Jakarta Faces Application*.

We will cover the following in this chapter:

- A brief history of Java GUIs
- A financial calculator program design
- Using the Swing GUI framework
- Using the JavaFX GUI framework
- Which should I use?

By the end of this chapter, you will understand the basics of GUI coding using the two most widely used frameworks.

Technical requirements

Here are the tools required to run the examples in this chapter:

- Java 17
- A text editor
- Maven 3.8.6 or a newer version installed

The sample code for this chapter is available at `https://github.com/PacktPublishing/Transitioning-to-Java/tree/chapter13`.

A brief history of Java GUIs

The first user interfaces in personal computers mimicked the terminals of mainframes or minicomputers. Apple introduced the Mac in 1984 and Microsoft introduced Windows a year later. Yet, most personal computers sold in the 1980s had terminal interfaces. What changed everything was the general availability of the internet and the creation of the technology behind the World Wide Web by Tim Berners-Lee, starting in 1989. By the end of the 20th century, we expected the computers we used to have a GUI. In *Chapter 15*, *Jakarta Faces Application*, we will look at web programming in Java, while in this chapter, we will look at desktop GUI programming.

When Java was introduced in 1995, its original purpose was the creation of applets, small programs that run from web pages inside a web browser. These pages delivered compiled applets that ran in the JVM rather than in the browser. JavaScript, also developed around the same time, ran inside the browser. This led to the first GUI library, which is still included in Java today, called the **Abstract Window Toolkit** (**AWT**), still usable for desktop apps. Applets are no more, having been deprecated.

The AWT depends on the underlying operating system for rendering a GUI. An AWT program running on an Apple Mac had a look and feel like a native Mac app. If you ran the same code on a Windows PC, its look and feel were like a native Windows app. As Java became popular in the application rather than web space, an enhanced GUI library, Swing – first introduced in 1996 – gained traction. Unlike AWT, Swing could render its own controls. Now, it was possible to develop a user interface that looked nearly identical on Windows, Mac, and Linux. In the *Further reading* list is a link to a Swing look-and-feel library called **Napkin**. It is an extreme example of how you could style Swing applications. Swing remains in wide use today. It is a standard library, included with all distributions of Java. It is maintained with bug fixes and minor enhancements.

In 2008, JavaFX 1.0, an alternative to Swing, was introduced. The version we will be looking at began as JavaFX 2.0 in 2011. The original purpose of JavaFX was to have a common platform for GUI apps on the desktop, the web, and mobile devices. Like Swing, it took care of its own rendering. It also introduced a declarative approach to defining a user interface, based on the XML language FXML. In this chapter, we will look at the imperative or coding approach to defining an interface.

Now, let us look at the application we will construct.

Financial calculator program design

In the previous chapter, we took the loan calculation that used doubles and changed it to use `BigDecimal`. We will continue to use this calculation plus two additional ones. One is a savings goal, whereby you indicate how much you wish to save, what the expected interest rate will be, and the number of months in which you wish to reach your goal. The second is the future value, whereby you can determine the amount of money you will have after saving the same amount for a specific number of months at an expected interest rate.

We will use the same data class and business class from the previous chapter. To the business class, we will add the two new calculations. We will not go over the new calculations in this chapter, as you can see them in the chapter's source code. For now, we will consider the GUI.

The application will present the user with three choices for the calculation. We want a single form into which we can enter the three values each calculation requires and where the result will appear. Finally, we want a button to signal that the calculation can be carried out. I usually work out the actual design on paper or a whiteboard, such as this figure:

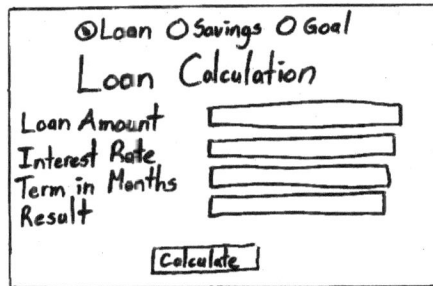

Figure 13.1 – A hand-drawn design

When we finish, it will look like this when doing a calculation:

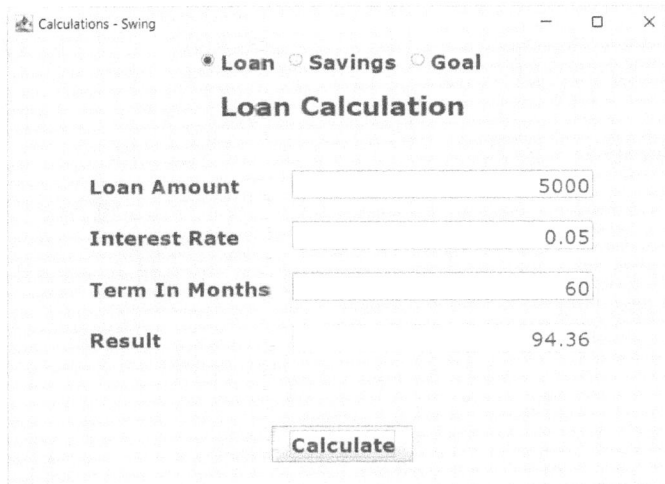

Figure 13.2 – The Swing version

The choice of calculation is determined by a group of three radio buttons. These are controls that are either selected or not, and you can only select one. When the user selects a calculation, the description of the title that follows the radio buttons will change, all fields will become zero, the first input field description will change, and pressing the **Calculate** button will use the calculation that matches the radio button selected.

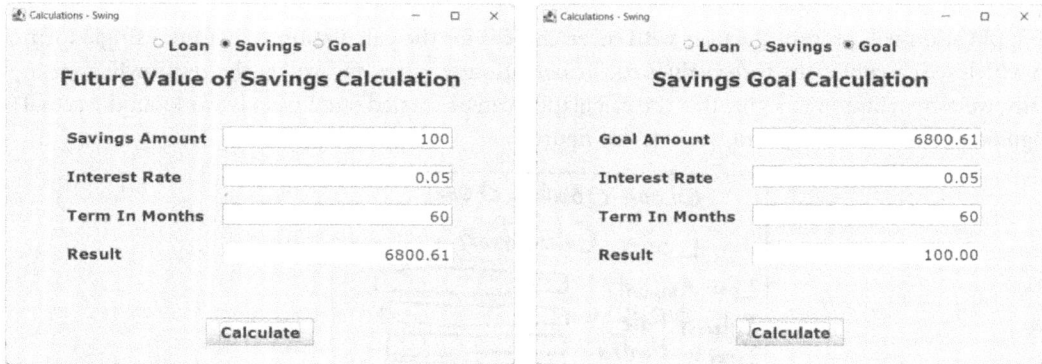

Figure 13.3 – The future value and savings goal screens

As you already know, users cannot always be trusted to enter valid information. There is always a user who will enter Bob for a loan amount. We could throw up message boxes informing the user of their error. In certain situations, this does make sense. In this user interface, we can recognize that the only allowable input in the first three fields must be a number and only allow a single decimal point. You just cannot enter Bob. A field that is blank will be changed to contain a zero. The last field is not editable, as it displays the result.

Internationalization – i18n

One last element of the design is internationalization, commonly referred to as i18n. This means that this program can present itself in multiple languages. I live in Canada, where we have two official languages – English and French. This means that anything that has written text in the interface must have a version in each language. Here it is in French:

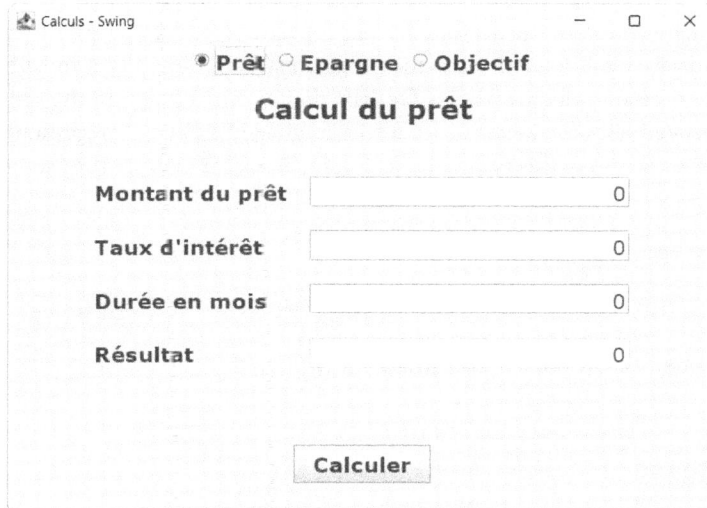

Figure 13.4 – The French version

We accomplish i18n by placing all text that will appear in the GUI into a properties file, one for each language you plan to support. Here is the English properties file:

```
loan_amount=Loan Amount
savings_amount=Savings Amount
goal_amount=Goal Amount
interest_rate=Interest Rate
term_in_months=Term In Months
result=Result
calculate=Calculate
calc_error=Arithmetic Exception
calc_msg=You may not have a zero in any input field.
title=Calculations - Swing
loan_title=Loan Calculation
savings_title=Future Value of Savings Calculation
goal_title=Savings Goal Calculation
loan_radio=Loan
savings_radio=Savings
goal_radio=Goal
alert_a=The value >%s< cannot \nbe converted to a number.
format_error=Number Format Error
```

You can see the French version in the project. These properties files are commonly called **message bundles**. The appropriate file is loaded when the program starts. All text that must be displayed is referenced by the key value in the properties file. The names of the keys are the same in bundles. Only the value associated with the key is in whatever language this bundle represents. This will be the same in the JavaFX version of this program. To employ `ResourceBundle`, you must first load the properties file.

Use the default `Locale`:

```
var form = ResourceBundle.getBundle("MessagesBundle");
```

Set `Locale` in the code, which can be quite useful for testing:

```
Locale locale = new Locale("en", "CA");
var form = ResourceBundle.getBundle(
                          "MessagesBundle", locale);
```

Internationalization works the same in JavaFX. Let us look at how this application is coded in Swing.

Using the Swing GUI framework

With the GUI library chosen, we can now decide what classes this application will need and how they will be packaged. Here is the finished project layout:

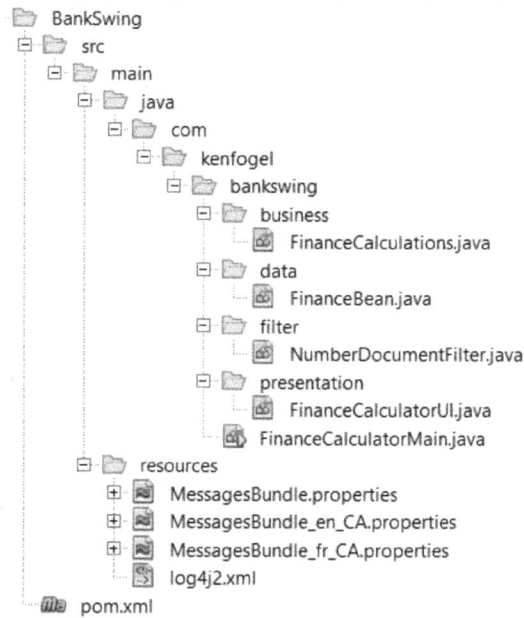

```
BankSwing
└── src
    └── main
        └── java
            └── com
                └── kenfogel
                    └── bankswing
                        └── business
                            └── FinanceCalculations.java
                        └── data
                            └── FinanceBean.java
                        └── filter
                            └── NumberDocumentFilter.java
                        └── presentation
                            └── FinanceCalculatorUI.java
                        └── FinanceCalculatorMain.java
        └── resources
            ├── MessagesBundle.properties
            ├── MessagesBundle_en_CA.properties
            ├── MessagesBundle_fr_CA.properties
            └── log4j2.xml
    └── pom.xml
```

Figure 13.5 – The Swing project layout

The source code on GitHub is extremely commented, and I encourage you to download it while reading this. Let us begin by looking at the basic components and controls of a Swing application. We begin with JFrame.

JFrame

Every Swing application needs a main or top-level container. There are four classes for this purpose, but one is now deprecated for removal. They are JFrame, JDialog, JWindow, and the deprecated JApplet. The JWindow class is ideal for splash screens, as they have no decorations, such as a border, title bar, or window controls. You commonly use JDialog as part of an application to interact with a user over details you do not want in the top-level container you are using. It can also be used for simple applications that require minimal interaction with the user and so can also be used as a top-level container. JApplet brought Swing to the web browser, but it is now headed to the dustbin of history. Let us talk about JFrame.

The JFrame class is a decorated container, meaning it has a border and title bar. It supports JMenuBar should you want a menu. It can be resizable or a fixed size. Into JFrame goes all other components and controls. In my sample code, I have used inheritance to extend JFrame in the FinanceCalculatorMain class. This simplifies the coding.

Here is the code that sets up the frame. You will see an additional step in the *JPanel* section:

```
setDefaultCloseOperation(JFrame.EXIT_ON_CLOSE);
setTitle(form.getString("title"));
setSize(620, 450);
setVisible(true);
```

JPanel

The JPanel class is a container for other components and controls. While the JFrame class already has the JPanel class, which is commonly referred to as the content pane, we rarely use it for anything more than adding a user-designed JPanel. This is exactly what my sample does. There is a second class called FinanceCalculatorUI that extends JPanel.

To add controls such as buttons, text fields, or other JPanels to a JPanel class, we must first decide on a LayoutManager class. These are objects that are responsible for the placement of items in JPanel. The main panel in the app is FinancialCalculatorUI, which extends JPanel. In its constructor, we write the following:

```
super(new BorderLayout());
```

Later in the code, another `JPanel` is created that does not extend another class. We can pass the layout manager through the `JPanel` constructor:

```
var innerForm = new JPanel(new BorderLayout());
```

It is quite common to create a user interface with multiple panels. The following diagram shows all the panels used:

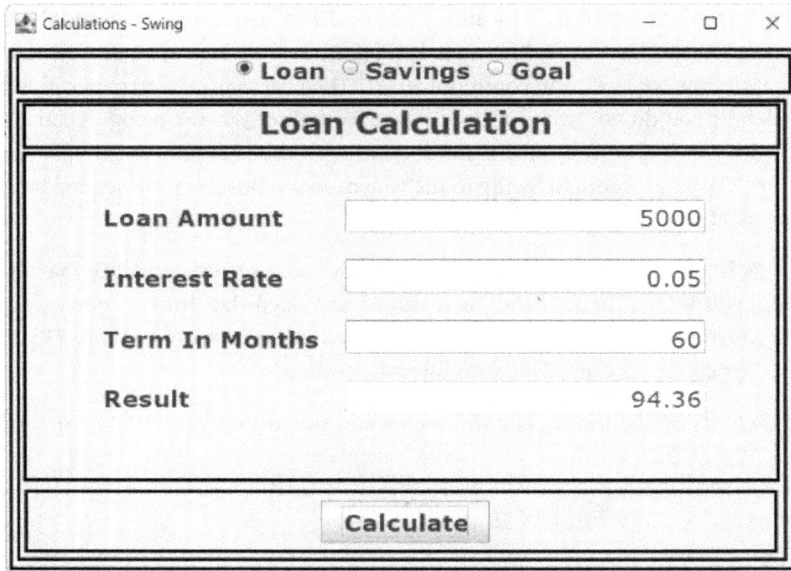

Figure 13.6 – The JPanel layout

The first `JPanel`, `FinancialCalculatorUI`, is assigned to the content pane of `JFrame`. It will have a `BorderLayout` that has north, south, east, west, and center zones. In each zone of a layout, you can add another panel or control. Anything you place in this layout will fill the zone. In the north, we placed a panel with `JRadioButtons`. In the center, we placed another `JPanel` with `BorderLayout`. In the north of this panel, we place a panel that contains the title. In the south, there is a panel with the calculate button. In the center, we place another panel with `GridBagLayout` that allows us to treat the panel as rows and columns, which is ideal for forms.

Event handlers

Any component that can accept user input from either the keyboard or mouse can generate an event. You can choose to ignore an event, or you can register an event handler you have written. Here is the code that registers a method in the same class as the handler:

```
calculate.addActionListener(this::calculateButtonHandler);
```

The `calculateButtonHandler` method accepts an `ActionEvent` object. The object contains information on the event, but in this case, we do not need it. The method validates that the fields can be `BigDecimal`, assigns them to `FinanceBean`, and, if all is well, calls on the `FinanceCalculations` object to do the calculation.

Document filter

In our design, we decided to only allow the end user to enter a number. This number can only have one decimal point and without a sign, because all values must be positive. In Swing, `JTextField` has inside it an object of type `Document`, and from this, we can access and replace the `DocumentFilter` object. In the filter, there is a method called `replace` that is called with every keystroke. We will install our own filter that inherits/extends `DocumentFilter`.

The `replace` method receives the string that you entered and the location where in the current text field this string is inserted. It is called before the text is entered in the document; we have access to the contents of the text field before it is changed. This means that should the user input turn out to be invalid, we can restore what was in the text field prior to the new string we entered.

Pattern matching with regular expressions

A *regular expression* or *regex* is a description of what may or may not be allowed in a string. Pattern matching employs a regex to determine whether a string that a pattern is being applied to meets the requirements of the regex. These expressions define a pattern that will be applied to the user input string:

```
private final Pattern numberPattern =
            Pattern.compile("^[\\d]*[\\.]?[\\d]*$");
private final Matcher numberMatch =
            numberPattern.matcher("");
```

The first line takes a regular expression and compiles it. Regular expressions are compiled every time you use them, if not compiled in advance. This is the role of creating a `Pattern` object.

The second line creates a `Matcher` object. Here, we create a `Matcher` object built from the `Pattern` object that can evaluate the string in the parenthesis and return `true` or `false`. In the `Matcher` declaration, it might seem strange to search in an empty string, but as we will see, you can change the string that you want to search for after you create the `Matcher` object:

```
if (numberMatch.reset(newText).matches()) {
    super.replace(fb, offset, length, string, attrs);
}
```

The reset method of Matcher is what allows us to change a string to search through. The newText object is the string after the new characters are added to the original string. If the match fails, then the string already in the text field is unchanged. The approach taken here will work if the user is pasting strings into the text field.

There are many developers who believe that pattern matching is costly, meaning it uses a lot of CPU time. Exceptions are also costly. If they are both costly, then the simpler exception approach will have the edge for most users. However, a simple test comparing the performance of pattern matching versus exception throwing, using the Java **Microbenchmark Harness** library, showed that pattern matching can be between 8 and 10 times faster than using an exception.

Controls and panels

Let us look at two controls that are placed in a panel with GridBagLayout. We begin with JLabel. We provide the text to display in its constructor. The actual text is coming from the appropriate ResourceBundle that the form object represents. We set the font next. This font is used in more than one place, so it has already been defined as labelFont. Finally, we add JLabel to the panel. To do this, we need an object of type GridBagConstraints. There are five details about how this control is added. The number 0 represents the column and the number 2 represents the row. The next two numbers represent how many columns and rows this control will use in the grid. Finally, we indicate the alignment of the control in the grid:

```
inputLabel = new JLabel(form.getString("loan_amount"));
inputLabel.setFont(labelFont);
panelForm.add(inputLabel,
    getConstraints(0, 2, 1, 1, GridBagConstraints.WEST));
```

The JTextField control has added settings for horizontal alignment and the width of the field measured in columns:

```
inputValue = new JTextField("0");
inputValue.setFont(textFont);
inputValue.setHorizontalAlignment(SwingConstants.RIGHT);
inputValue.setColumns(15);
```

Here, we extract the control's Document and use this reference to install our filter:

```
var inputDocument =
    (AbstractDocument) inputValue.getDocument();
inputDocument.setDocumentFilter(filter);
```

Finally, we add JTextField to the grid:

```
panelForm.add(inputValue,
    getConstraints(1, 2. 1, 1, GridBagConstraints.WEST));
```

You will find more details on this Swing version in the source code. Now, let us look at the JavaFX version.

Using the JavaFX GUI framework

This version of the program is like the Swing version. The design of the user interface is identical in that it employs panes in panes. Here is the finished project layout:

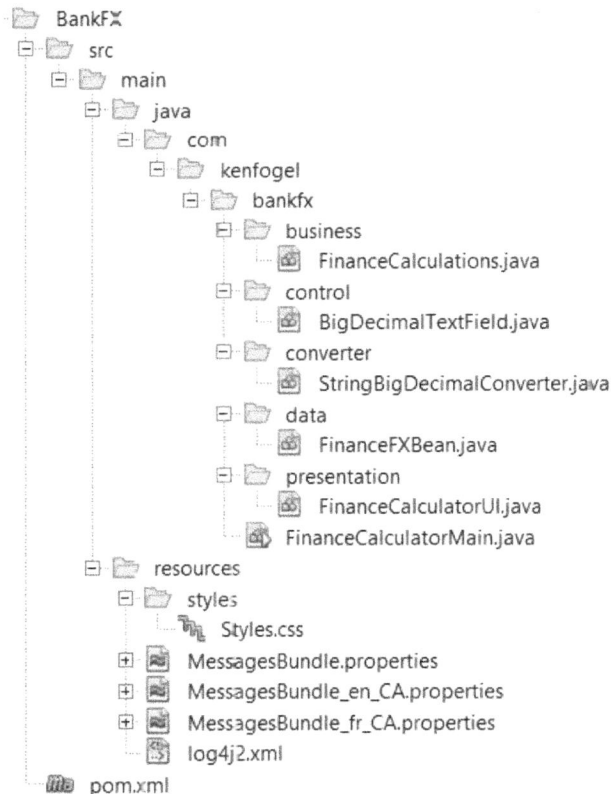

Figure 13.7 – The JavaFX program layout

Let us now look at the classes from the JavaFX framework that we will need for our program.

Application

A JavaFX program must contain a class that extends `Application`. Within this class, we can construct the user interface or delegate this work to another class. A class that extends `Application` must implement a method called `start` and, optionally, a method called `init`. What you rarely have is a constructor. The JavaFX framework is not available to a constructor of a class that extends `Application`. This is where `init` comes in. It plays the role of the constructor but in an environment where JavaFX is up and running. You do not call `init`; JavaFX will.

The `start` method is where the creation of the GUI commences. The method is called by JavaFX right after `init`.

PrimaryStage

The `PrimaryStage` object is akin to `JFrame`. You do not create an instance of it. The `Application` class you are extending creates a `PrimaryStage` object and passes it on to `start`:

```
@Override
public void start(Stage primaryStage) {
```

Pane

Unlike Swing, where you define a panel and assign a layout manager to it, FX uses panes that include the layout. The first pane that we are creating is `BorderPane`, which has top, bottom, left, right, and center zones. We are adding to it a pane that contains the radio buttons at the top and another pane in the center:

```
var root = new BorderPane();
root.setTop(gui.buildRadioButtonsBox());
root.setCenter(gui.buildForm());
```

Scene

The story goes that the original developer of JavaFX was a performing arts aficionado, and this led him to use theater names for parts of the framework. A `Stage` object must contain a `Scene` object that, in turn, contains a pane object. You can create multiple scenes and switch between them if required. For me, one of the most noteworthy features of JavaFX is its use of **Cascading Style Sheet (CSS)**, and the second line will load it.

```
var scene = new Scene(root, 620, 450);
scene.getStylesheets().add("styles/Styles.css");
```

The next four lines should be self-explanatory. The last line calls the show method to get the program going

```
        primaryStage.setTitle(form.getString("title"));
        primaryStage.setScene(scene);
        primaryStage.centerOnScreen();
        primaryStage.show();
    }
```

CSS style sheets

You style components using CSS. This is similar but not identical to CSS used in web development. In the *Further reading* section, there is a link to the CSS reference document. Here is the style sheet used in this application. Class names beginning with a period are predefined. Class names beginning with an octothorpe are the ones that you assign:

```
#prompt_label {
    -fx-font-size:14pt;
    -fx-font-weight:bold;
    -fx-font-family:Verdana, sans-serif;
}
#input_field {
    -fx-font-size:14pt;
    -fx-font-weight:normal;
    -fx-font-family:Verdana, sans-serif;
}
#title {
    -fx-font-size:18pt;
    -fx-font-weight:bold;
    -fx-font-family:Verdana, sans-serif
}
.button {
    -fx-font-family:Verdana, sans-serif;
    -fx-font-weight:bold;
    -fx-font-size:14pt;
}
.radio-button {
    -fx-font-family:Verdana, sans-serif;
    -fx-font-weight:bold;
```

```
            -fx-font-size:14pt;
    }
```

You can assign a custom class name for a control by using the control's `setId` method:

```
    resultLabel.setId("prompt_label");
```

You can also enter the CSS styling directly in the source code. CSS entered this way will override what is in the external style sheet:

```
    resultLabel.setStyle("-fx-font-size:18pt; "
        + "-fx-font-weight:bold; "
        + "-fx-font-family:Verdana, sans-serif;");
```

JavaFX bean

A JavaFX bean is designed to support binding a field in the bean with a control. Any change to the control is written to the bean, and anything written to the bean will update the control. This is commonly called the observer pattern. To accomplish this, we must wrap all the data types into a family of objects called properties. A JavaFX bean can be used where a JavaBean is expected. This means that there is no need to alter the `FinancialCalculations` class, as it will work with the JavaFX bean without any changes to its code. Here is our `FinanceFXBean` class as a JavaFX bean:

```
    public class FinanceFXBean {
```

Rather than declare primitive or class types directly, they must be wrapped into an appropriate `Property` class. There are such classes for all the primitive data types, such as `DoubleProperty` or `StringProperty`. In our example, we are using `BigDecimal`, for which there is no property. This is where the `ObjectProperty` class comes in. This allows you to use any class as a property:

```
    private ObjectProperty<BigDecimal> inputValue;
    private ObjectProperty<BigDecimal> rate;
    private ObjectProperty<BigDecimal> term;
    private ObjectProperty<BigDecimal> result;
```

There are two constructors. The first is a default constructor that calls upon the non-default constructor:

```
    public FinanceFXBean() {
        this(BigDecimal.ZERO, BigDecimal.ZERO, BigDecimal.ZERO);
    }
```

The non-default constructor must instantiate each of the `Property` objects and initialize them, with `BigDecimal` passed to it. There is a simple property for each of the defined properties. `SimpleObjectProperty` is used when there is not a named property, such as `SimpleDoubleProperty`:

```
public FinanceFXBean(BigDecimal inputValue,
                BigDecimal rate, BigDecimal term) {
    this.inputValue =
            new SimpleObjectProperty<>(inputValue);
    this.rate =
            new SimpleObjectProperty<>(rate);
    this.term =
            new SimpleObjectProperty<>(term);
    this.result =
            new SimpleObjectProperty<>(BigDecimal.ZERO);
}
```

Unlike a JavaBean, there are three methods for each property. The first two, the getter and setter, retrieve the value from inside the property or change the value in the property. This is how a JavaFX bean can replace a Java bean, as the getter and setters are named the same and return or receive the same data types:

```
public BigDecimal getInputValue() {
    return inputValue.get();
}
public void setInputValue(BigDecimal inputValue) {
    this.inputValue.set(inputValue);
}
```

The last of the three methods returns the property. This is used to implement binding and the observable pattern:

```
public ObjectProperty<BigDecimal> inputValueProperty() {
    return inputValue;
}
```

BigDecimalTextField

This is a class that extends the JavaFX `TextField`. Unlike Swing's `JTextField`, there is no `Document`. Pressing a key will invoke a method in `TextField` that we override to meet our needs. These methods use regular expressions for pattern matching. There is now a second regular expression that is used to modify the string by removing leading zeros.

There are two methods that we override. The first is `replaceText` for handling keyboard input. The second is `replaceSelection` for handling changes that arise from selecting text and then replacing it, either from the keyboard or by pasting.

Controls

There is a similarity in how we create controls and add them to a pane. You create `GridPane` to hold the form:

```
var loanGrid = new GridPane();
```

`inputLabel` is instantiated using `ResourceBundle`. A custom CSS-style `id` is assigned next. Finally, we add the label to the `financeGrid` pane, indicating the column and row:

```
inputLabel = new Label(form.getString("loan_amount"));
inputLabel.setId("prompt_label");
financeGrid.add(inputLabel, 0, 0);
```

`inputValue` is constructed from our custom `BigDecimalTextField`. We assign it a custom CSS `id`, change its alignment, and add it to `financeGrid`:

```
inputValue = new BigDecimalTextField();
inputValue.setId("input_field");
inputValue.setAlignment(Pos.CENTER_RIGHT);
financeGrid.add(inputValue, 1, 0);
```

Binding

Binding a primitive data type to a field in a JavaFX bean requires a converter. There are some standard converters, one of which is for `BigDecimal` to convert from a string to a BigDecimal. Here is the binding method:

```
private void doBinding() {
```

We are creating a `BigDecimalStringConverter` object that we can reuse in each of the bindings:

```
var convert = new BigDecimalStringConverter();
```

For each binding, we pass `textProperty` of each `TextField`, the property for each field in the `FinanceFXBean`, and our converter:

```
Bindings.bindBidirectional(inputValue.textProperty(),
    financeData.inputValueProperty(), convert);
```

```
    Bindings.bindBidirectional(rateValue.textProperty(),
        financeData.rateProperty(), convert);
    Bindings.bindBidirectional(termValue.textProperty(),
        financeData.termProperty(), convert);
    Bindings.bindBidirectional(resultValue.textProperty(),
        financeData.resultProperty(), convert);
}
```

Which should you use? Read the following summary to decide.

Summary

Swing is a mature GUI framework that is maintained as part of the Java distribution. Having been around for so long, there has been much written on how to use it. Choosing Swing for your GUI project is a good choice.

JavaFX is the new kid on the block, so to speak. The two most significant differences are binding and CSS style sheets. Although not covered here, JavaFX also has much better graphics support. In JavaFX graphics, primitives such as a line or a rectangle are on par with controls. JavaFX has an Animation class that simplifies creating moving graphics. There is a also chart library to create line charts, pie charts, and so on.

Users today expect a GUI in the software they use. Swing projects do not need to be rewritten using JavaFX, and new projects using Swing continue to be written. So, which should you use? In my opinion, new projects should use JavaFX. Swing is, for the most part, in maintenance mode to ensure that it works, taking advantage of changes in the core language. JavaFX, on the other hand, is actively maintained as an open source project, with both maintenance and new features added to each release.

In our next chapter, we will look at how Java is used to write software for the World Wide Web that runs in a container called an application server.

Further reading

- Napkin look and feel: `https://napkinlaf.sourceforge.net/`

- Java Microbenchmark Harness: `https://medium.com/javarevisited/understanding-java-microbenchmark-harness-or-jmh-tool-5b9b90ccbe8d`

- Pattern matching: `https://xperti.io/blogs/pattern-matching-java-feature-spotlight/`

- *JavaFX CSS Reference Guide*: `https://openjfx.io/javadoc/17/javafx.graphics/javafx/scene/doc-files/cssref.html`

14
Server-Side Coding with Jakarta

While Java's initial connection to the web was through applet development, it was only a few years after the language appeared that server-side Java, first called the **Java 2 Enterprise Edition**, or J2EE, and later called the **Java Enterprise Edition**, or **JEE**, was introduced. Unlike standalone applications that can run on your desktop, JEE applications run inside another family of Java programs called application servers. When Oracle decided to focus primarily on the core language, Java SE, the specifications and libraries were turned over to the Eclipse Foundation. These specifications and libraries were renamed **Jakarta EE**.

Server-side coding in any programming language typically involves software listening to an internet port, such as 80. The passing of information from a browser to a server and back again follows the HTTP protocol. A browser delivers a request to a server. The request may be satisfied by returning a response that consists of HTML and JavaScript to the browser, which, in turn, renders a page. The HTTP protocol is language- and server-agnostic. This means that it is not tied to a specific language or browser.

In Java, we have a special type of class that runs in the application server waiting for a request, performs some action when the request is received, and then returns the response that the browser can render. This special class is a **servlet**. A servlet is called upon as if it were a regular web page. When a request arrives at the application server that must be fulfilled by a servlet, a thread to the servlet is created by the server. This means that should 100 users request the same servlet, each request gets a thread, rather than instantiating the entire object for each request. Threads are faster to create and have a smaller memory footprint than an ordinary object created with the new keyword.

A Jakarta EE application is packaged in a ZIP file with a `.war` extension. It does not include the JEE libraries. These libraries are part of the application server. This means that a web app is relatively small. These libraries are required to compile the code. Maven will bring down the libraries so that the Java compiler can validate your usage of Jakarta. When packaging an app into a `.war` file for the server, these libraries are not part of the final package.

The Jakarta EE specification describes two page-rendering libraries. The first is called **Jakarta Server Pages**, previously **JavaServer Pages**, or **JSP**. The second is called **Jakarta Faces**, previously known as **JavaServer Faces**, or **JSF**. This acronym is still used widely, rather than JF. Both libraries support the generation of HTML and JavaScript from Java code running in the application server. We will look at Jakarta Faces in the next chapter.

In this chapter, we will look at the following topics:

- Understanding the role of the Java application server

- Configuring a web project with Maven

- Understanding what a servlet does and how it is coded

- Configuring deployment with the web.xml file

By the end of this chapter, you should be able to understand how a web application is constructed in Java based on HTML and how servlets are coded.

Technical requirements

Here are the tools required to run the examples in this chapter:

- Java 17

- A text editor

- Maven 3.8.6 or a newer version installed

- The GlassFish 7.0 application server

- A web browser

The sample code for this chapter is available at https://github.com/PacktPublishing/Transitioning-to-Java/tree/chapter14.

Understanding the role of the Java application server

The center of the Jakarta EE 10 programming universe is the application server. These programs provide a range of services that your application can call upon. They also contain all the Jakarta 10 libraries that your application might need. This simply means that your application does not need to include all the required external libraries, such as what a desktop application must include in the final JAR file.

An application server can be designated in one of three ways:

- The first is the platform. It provides the entire set of Jakarta EE 10 services.

- The second is the Web profile, which provides a subset of the platform services.

- Finally, there is the Core profile. The smallest of the profiles, it is designed to provide the infrastructure for microservices.

The following table shows which Jakarta EE 10 libraries can be found in each profile. Libraries in the columns to the right of each profile, except for Core, are in that profile. The platform includes the Web profile and the Core profile, while the Web profile includes the Core profile. As Jakarta EE evolves, new features can be added and libraries are updated.

Platform	Web profile	Core profile
Authorization 2.1	Expression Language 5.0	CDI Lite 4.0
Activation 2.1	Authentication 3.0	JSON Binding 3.0
Batch 2.1	Concurrency 3.0	Annotations 2.1
Connectors 2.1	Persistence 3.1	Interceptors 2.1
Mail 2.1	Faces 4.0	RESTful Web Services 3.1
Messaging 3.1	Security 3.0	JSON Processing 2.1
Enterprise Beans 4.0	Servlet 6.0	Dependency Injection 2.0
	Standard Tag Libraries 3.0	
	Server Pages 3.1	
	CDI 4.0	
	WebSocket 2.1	
	Bean Validation 3.0	
	Debugging Support 2.0	
	Enterprise Beans Lite 4.0	
	Managed Beans 2.0	
	Transactions 2.0	

Table 14.1 – Jakarta EE 10 libraries/services

A platform server is expected to provide all the services listed in the previous table. A Web profile server provides Web profile and Core profile services. Finally, a Core profile server only supports what is in its column. We will only look at a few of these services in this chapter.

Application servers are available from several companies. These servers usually have a free community/ open source edition, as well as versions with paid licensing. Paid licensing gets you support for the server. Community editions maintain mailing lists, on which you can ask questions and get a response from either the company or other users of the community editions. One specific server stands out, and that is open source. This is the Eclipse GlassFish server. This is the one we will use in this chapter.

GlassFish 7.0

The GlassFish server was initially developed by Sun Microsystems as the reference server for Java EE. This meant that any other server that wished to be identified as a Java EE server needed to pass the same **Technical Compatibility Kit** test suite, commonly called the **TCK**, as GlassFish.

When Oracle acquired Sun, they continued to maintain GlassFish. In 2017, Oracle decided to no longer develop Java EE. They designated the Eclipse Foundation as the new home for Java EE, who, in turn, renamed it Jakarta EE. The technology transfer included GlassFish. This also meant that Jakarta EE and GlassFish are open source.

Downloading, installing, and running GlassFish

The GlassFish server can be downloaded from https://glassfish.org/. There are two choices for a standalone server. These are as follows:

- Eclipse GlassFish 7.0.0 and Jakarta EE Platform 10
- Eclipse GlassFish 7.0.0 and Jakarta EE Web Profile 10

In addition, there are two embedded versions. An embedded version can be used as part of an application. There is just one download for each choice. These are not Linux, macOS, or Windows versions, as they all use nearly the same class files and libraries, and any specific components for a given OS are part of the single version. It is a ZIP file. Installation is quite simple. Here are the steps:

1. Download GlassFish.
2. Set the environment or JAVA_HOME shell variable to the location of your JVM.
3. Unzip the file you downloaded. It should create a folder called glassfish7 that you can now move to where you wish it to be.
4. Go to the bin folder in the glassfish7 folder.
5. Open a terminal or console window in the bin folder.
6. Start the server for any OS by entering asadmin start-domain.

   On Linux, ensure that the asadmin shell script is executable before you run it. On Windows, you will be running the asadmin.bat batch file. To stop a server, enter asadmin stop-domain. You should see messages in the console/terminal window, telling you that you are successful. If you are not, then please review the more detailed installation instructions on the GlassFish website.

7. To test the installation, open your web browser and enter http://localhost:8080.

   The default ports that GlassFish listens to are 8080 for applications running on the server and 4848 for access to the admin console. If needed, both these ports can be changed.

You should see a web page that looks like this:

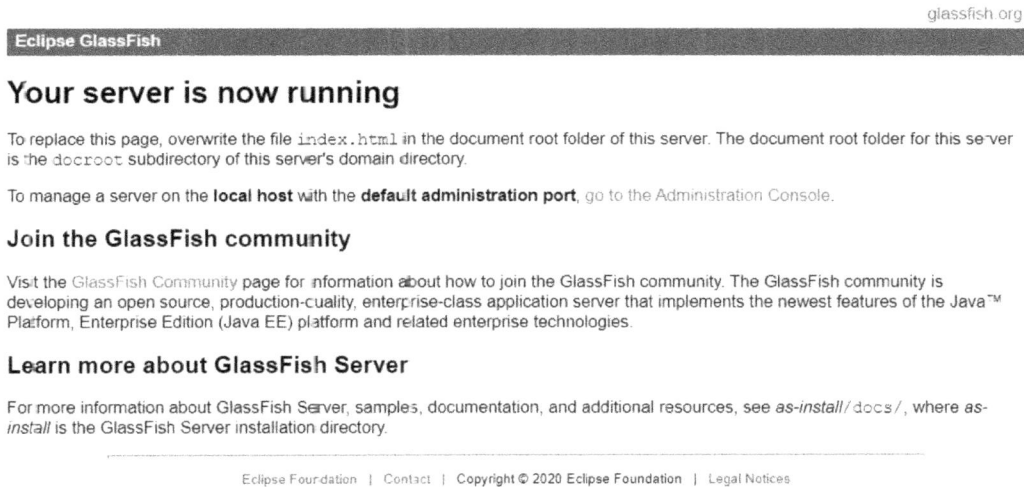

Eclipse GlassFish

Your server is now running

To replace this page, overwrite the file index.html in the document root folder of this server. The document root folder for this server is the docroot subdirectory of this server's domain directory.

To manage a server on the **local host** with the **default administration port**, go to the Administration Console.

Join the GlassFish community

Visit the GlassFish Community page for information about how to join the GlassFish community. The GlassFish community is developing an open source, production-quality, enterprise-class application server that implements the newest features of the Java™ Platform, Enterprise Edition (Java EE) platform and related enterprise technologies.

Learn more about GlassFish Server

For more information about GlassFish Server, samples, documentation, and additional resources, see *as-install*/docs/, where *as-install* is the GlassFish Server installation directory.

Eclipse Foundation | Contact | Copyright © 2020 Eclipse Foundation | Legal Notices

Figure 14.1 – The default port 8080 web page

8. To access the admin console, enter http://localhost:4848.

You should now see the following:

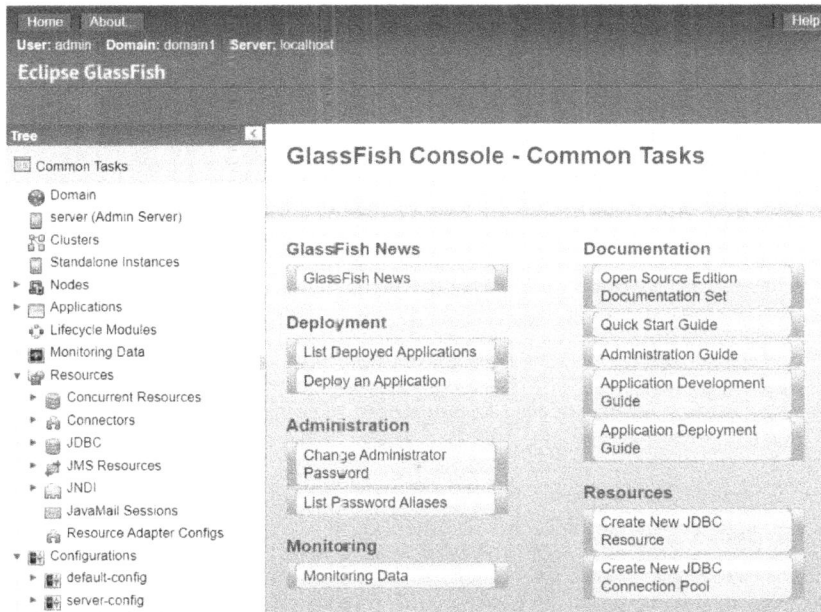

Figure 14.2 – The port 4848 admin console

You should use **Change Administrator Password** under **Common Tasks** to assign a password for the admin user, as you likely noticed that you were not asked for a password to access the admin console. For our purposes, there is nothing more to be done with GlassFish. When we wish to test our applications, we can use the **Deploy an Application** common task.

Explore GlassFish and read its documentation, which you will find on the download site. Its default configuration is all we need. Let us now create the necessary folder structure to build a web app with Maven.

Configuring a web project with Maven

The first step to crafting a web app is to configure your project for Maven. First, we need to create the appropriate folder structure for any Jakarta EE application built with Maven. Here is what you need to set it up:

Figure 14.3 – The required folders for a Jakarta EE app being built with Maven

The only difference between a Maven desktop setup and a web setup is the addition of the webapp folder in the main folder. In this folder is a WEB-INF folder and an optional styles folder. Here is a rundown of the folder:

- src/main/java: All Java source files are stored in subfolders/packages, just as we did in a desktop app.
- src/main/resources/: Language bundles and logger configuration files go here. Some bundles can be placed in subfolders while others cannot.
- src/main/webapp: This is the folder that will contain any static web pages, JavaServer pages, and JavaServer Faces pages. You can create subfolders.

- `src/main/webapp/WEB-INF`: This folder contains configuration files and private files. A private file can be anything that might just be in the `WEB-INF` folder. A URL cannot include this folder, and this is why they are considered private. The folder can be accessed by code running on the server.

- `src/main/webapp/styles`: This folder will hold any CSS files. This is not a standard folder, so you could place your CSS files in any folder, except `WEB_INF` in the `webapp` folder.

- `src/test`: This is the home of any unit tests or other files used exclusively when running unit tests.

When your code is ready to be compiled, you only need to open a terminal/console window in the project's root folder and enter the Maven command, `mvn`. If there were no errors, then you will have a new folder called `target` in your project, and in here, you will find the `.war` file. A `.war` file, like a `.jar` file, is a ZIP compressed file. The difference between them is how they are laid out in the file. Web servers expect an organization of files that is different from a desktop program.

Changes to the pom.xml file

A web application is packaged in a file with the `war` extension. The folder organization in this file is based on the standard for application servers. This means that the first change to the POM file will be as follows:

```
<packaging>war</packaging>
```

In our desktop pom file, we included dependencies for logging and unit testing. We will use `java.util.logging`, thus eliminating all the logging dependencies. Unit testing for web applications requires a special code runner, such as Arquillian from Red Hat. We will not be covering this and, therefore, can remove the unit testing dependencies and plugins. The new `pom.xml` file will now, starting with `properties`, contain the following:

```
<properties>
    <java.version>17</java.version>
    <project.build.sourceEncoding>
        UTF-8
    </project.build.sourceEncoding>
    <maven.compiler.release>
        ${java.version}
    </maven.compiler.release>
    <jakartaee>10.0.0</jakartaee>
</properties>
```

In the `dependencies` section that follows, we show the Jakarta library dependency. Note that the `scope` setting is set to `provided`, which implies that the libraries do not get included in the WAR file:

```
<dependencies>
    <dependency>
        <groupId>jakarta.platform</groupId>
        <artifactId>jakarta.jakartaee-api</artifactId>
        <version>${jakartaee}</version>
        <scope>provided</scope>
    </dependency>
</dependencies>
<build>
    <finalName>
        ${project.artifactId}
    </finalName>
```

Unlike desktop applications, we cannot simply run a web app. It must be copied to the appropriate folder in GlassFish and then a browser opens the site. While there are Maven plugins that can do this for you, we will keep it simple. Maven will output a WAR file, and you can use the GlassFish admin console to deploy it:

```
<defaultGoal>verify package</defaultGoal>
<plugins>
    <plugin>
        <groupId>org.apache.maven.plugins</groupId>
        <artifactId>maven-war-plugin</artifactId>
        <version>3.3.2</version>
    </plugin>
</plugins>
</build>
```

There is another way to deploy an app. There is a folder in GlassFish called `glassfish7\glassfish\domains\domain1\autodeploy`.

If you simply copy a WAR file to this folder, then the server will deploy it automatically.

Now, let's look at the heart of Java web programming, the servlet.

Understanding what a servlet does and how it is coded

In Java web programming, there is no main method. Instead, all applications must have at least one servlet. When we look at Jakarta Faces' client-side rendering, there is no servlet, as it is already part of the library. Let's look at a servlet.

The first line is an annotation that defines that this class is a servlet. The description is visible in the server's admin console. The urlPattern attribute is the name you use in a URL. A servlet can be named anything and can have any extension, although the standard practice is to not have an extension. A servlet can have multiple patterns. Here is an example of a servlet:

```
@WebServlet(description = "Basic Servlet",
                  urlPatterns = {"/basicservlet"})
```

If we wished to refer to this servlet with more than one pattern, we could write the following:

```
@WebServlet(description = "Basic Servlet",
                  urlPatterns = {"/basicservlet",
                                      "/anotherservlet"})
```

A servlet in Java is a class that extends HttpServlet:

```
public class BasicServlet extends HttpServlet {
    private static final Logger LOG =
           Logger.getLogger(BasicServlet.class.getName());
```

The constructor of a servlet class is rarely used because it cannot call upon any methods in the HttpServlet superclass. You can safely leave it out:

```
    public BasicServlet() {
        LOG.info(">>> Constructor <<<");
    }
```

If you must prepare for or initialize something before the servlet receives its first request, then you can use the init method. It can access the superclass, but it does not get a request or response object. It is called when the servlet is run before the first request arrives:

```
    @Override
    public void init() throws ServletException {
        LOG.info(">>> init <<<");
    }
```

The `destroy` method is akin to a destructor in C++. It is called by the server before it is unloaded to carry out any required end-of-life tasks:

```
@Override
public void destroy() {
    LOG.info(">>> destroy <<<");
}
```

The `getServletInfo` method allows you to prepare a string with information on this servlet:

```
@Override
public String getServletInfo() {
    LOG.info(">>> getServletInfo <<<");
    return "BasicServlet01 Version 2.0";
}
```

The `service` method is called by the server whenever a request is made to this servlet. The `service` method calls upon the `HttpServletRequest` object's `getMethod` to determine the type of request and then calls the matching do method, such as `doPost` or `doGet`. The most common reason for overriding this is if you wish to perform a task regardless of the request type. In this example, we are just calling the `service` superclass method that you must do if you are not calling the appropriate method in the body of the overridden `service` method:

```
@Override
protected void service(HttpServletRequest request,
        HttpServletResponse response)
        throws ServletException, IOException {
    super.service(request, response);
    LOG.info(">>> service <<<");
}
```

There are eight different types of requests. These are the HTTP verbs that the servlet provides to support the protocol. They are GET, POST, PUT, DELETE, HEAD, OPTIONS, CONNECT, and TRACE. The first four are the most used, although only GET and POST can be used on an HTML page. To test requests that cannot be issued from an HTML page, you can use the `curl` utility. This tool allows you to send any type of request from the terminal/console of your computer. When you run this application, you will see instructions for downloading and using `curl`:

```
@Override
protected void doGet(HttpServletRequest request,
```

```
                    HttpServletResponse response)
            throws ServletException, IOException {
```

There are different content types that can be returned to a browser in the `response` object. The type could be, among others, `image/gif` or `application/pdf`. Plain text is `text/plain`:

```
        response.setContentType("text/html;charset=UTF-8");
```

For a servlet to return text to a browser we use a `PrintWriter` object. It is instantiated by the `response` object such that the data you are writing will go to the URL found in the `response` object:

```
        try ( PrintWriter writer = response.getWriter()) {
            writer.print(createHTMLString("GET"));
        }
        LOG.info(">>> doGet <<<");
    }
```

Here is the `doPost` method that will display the web page created by `createHTMLString` and show that a POST request is issued:

```
    @Override
    protected void doPost(HttpServletRequest request,
            HttpServletResponse response) throws
            ServletException, IOException {
        response.setContentType("text/html;charset=UTF-8");
        try ( PrintWriter writer = response.getWriter()) {
            writer.print(createHTMLString("POST"));
        }
        LOG.info(">>> doPost <<<");
    }
```

Here is the `doPut` method. As we can only issue PUT using `curl`, all it returns is a simple string that `curl` will display in your terminal/console:

```
    @Override
    protected void doPut(HttpServletRequest request,
            HttpServletResponse response)
            throws ServletException, IOException {
        response.setContentType(
```

```
                              "text/plain;charset=UTF-8");
        try ( PrintWriter writer = response.getWriter()) {
            writer.print("You have called doPut");
        }
        LOG.info(">>> doPut <<<");
    }
```

Here is the doDelete method. Just as with PUT, you can only issue it using curl:

```
    @Override
    protected void doDelete(HttpServletRequest request,
            HttpServletResponse response)
            throws ServletException, IOException {
        response.setContentType(
                        "text/plain;charset=UTF-8");
        try ( PrintWriter writer = response.getWriter()) {
            writer.print("You have called doDelete");
        }
        LOG.info(">>> doDelete <<<");
    }
```

This last method is a user method that is used to generate a string of HTML code, which can be returned to the user's browser. Note that the HTML page is enclosed in a text block using the three quotation marks. There is also a placeholder, %s, in the text that is replaced using the formatted string method:

```
    private String createHTMLString(String methodName) {
        String htmlStr = """
            <html>
                <head><link rel='stylesheet'
                    href='styles/main.css'
                    type='text/css'/>
                <title>The Basic Servlet</title></head>
                <body>
                    <h1>%s method</h1>
                    <br/><br/>
                    <form action='index.html'>
                        <label>Return to Home page</label>
```

```
                <br/>
                <button class='button'>Return</button>
            </form>
        </body>
    </html>
    """.formatted(methodName);
        return htmlStr;
    }
}
```

What happens when a servlet is requested?

A servlet class is instantiated by the server either when the server begins or when the servlet is called for the first time. Once instantiated, it remains in the server until you explicitly ask the server to remove it. There is only one instance of every servlet.

Each request generates a thread of the servlet. Creating threads is faster than creating objects. The thread of the servlet is free to do almost anything it wants, such as instantiating other objects. Should a thread not receive a request within a user-defined time period, usually 30 minutes, it is stopped and the objects created by the thread go to garbage collection.

How does a servlet access the query string in a request?

Assume an HTML form that has three input fields named emailAddress, firstName, and lastName. Clicking on a button of the submit type will create a query string that will be appended to the URL if you are using a GET request, or added to the request body if you are using POST. In both cases, the data is in the key = value format. Here is such an HTML page:

```
<html>
    <head>
        <title>Just Servlet Input</title>
        <link rel="stylesheet" href="styles/main.css"
                                    type="text/css"/>
    </head>
    <body>
        <h1>Join our email list</h1>
        <p>To join our email list, enter your name and
            email address below.</p>
```

```
<form action="AddToEmailList" method="get">
    <label class="pad_top">Email:</label>
    <input type="email" name="emailAddress"
                                required><br>
    <label class="pad_top">First Name:</label>
    <input type="text" name="firstName"
                                required><br>
    <label class="pad_top">Last Name:</label>
    <input type="text" name="lastName"
                                required><br>
    <label> </label>
    <input type="submit" value="Join Now"
                            class="margin_left">
</form>
</body>
</html>
```

This HTML will produce the following page:

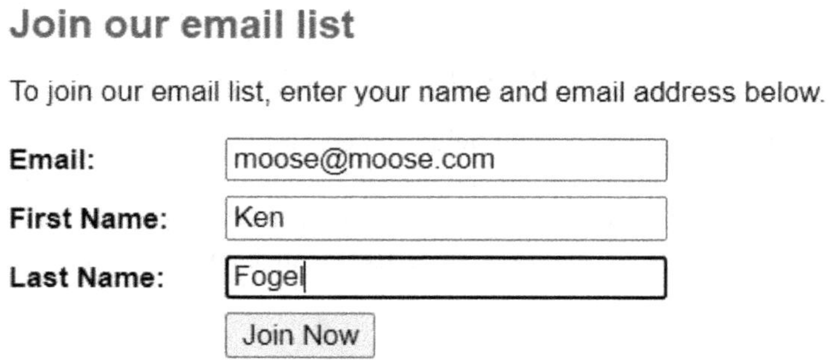

Join our email list

To join our email list, enter your name and email address below.

Email:	moose@moose.com
First Name:	Ken
Last Name:	Fogel

Join Now

Figure 14.4 – The browser rendering of the HTML

In HTML, I use the method attribute to show that the type of request issued when the button is pressed is GET. As this form submits data to the server, it should use the POST method. I use GET here, as it shows the query string in the address bar, whereas POST transmits the query string in a separate component of a request and, therefore, is not visible. POST should also be preferred should you need to prevent the information in the query string from being sent as plain text and also shown in the server logs.

I have already filled out the form, and when I click on the button, the URL in the browser will be updated to show as a single line:

```
http://localhost:8080/HTMLServlet/AddToEmailList?
                            emailAddress=moose%40moose.com&
                            firstName=Ken&lastName=Fogel
```

The doGet method in the servlet can now read the three parameters. In my example, I am storing this data in a simple JavaBean-style object:

```
@Override
protected void doGet(HttpServletRequest request,
            HttpServletResponse response)
            throws ServletException, IOException {
```

Using the names of the key values in the query string, we can retrieve the data and then assign them to the User object:

```
String firstName =
        request.getParameter("firstName");
String lastName = request.getParameter("lastName");
String emailAddress =
        request.getParameter("emailAddress");

User user =
        new User(firstName, lastName, emailAddress);
```

Here, I am displaying a results page constructed in a method called displayConfirmation:

```
response.setContentType("text/html;charset=UTF-8");
try (PrintWriter out = response.getWriter()) {
    displayConfirmation(out, user);
}
}
```

The servlet's output will be as follows:

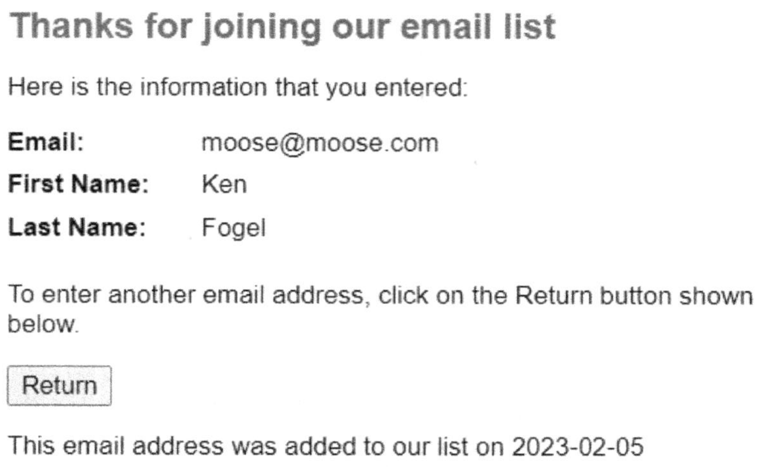

Thanks for joining our email list

Here is the information that you entered:

Email: moose@moose.com
First Name: Ken
Last Name: Fogel

To enter another email address, click on the Return button shown below.

Return

This email address was added to our list on 2023-02-05

Figure 14.5 – Output from the servlet

Do not forget to review the source code for this chapter.

How does a servlet remember my data?

Every time you call upon a servlet that is part of an application for the first time, you will receive an ID number that identifies an `HttpSession` object. This ID is sent as a cookie to your browser or, if you are blocking cookies, as a hidden field in the URL every time a request is made. If you already have the ID in a cookie, then that will be used. The server manages the ID; you do not need to do anything. The server uses this ID to manage the `HttpSession` object and ensure that your requests are the only ones that have access. You access the session object with the following code:

```
HttpSession session = request.getSession();
```

If an `HttpSession` object associated with your ID exists, then it is returned. If it does not exist, then a new `HttpSession` object with its own ID is returned. We will use one of two methods in the `HttpSession` object, one for reading and one for writing. If you wanted to preserve the `User` object in this example so that it can be used in another servlet, you will code the following:

```
HttpSession session = request.getSession();
session.setAttribute("myUser", user);
```

This `HttpSession` object will remain valid until the `HttpSession` timer, usually 30 minutes, ends or you call `session.invalidate()`. If we want to retrieve the `User` object in another servlet, then we can write the following:

```
HttpSession session = request.getSession();
String animal = (User) session.getAttribute("myUser");
```

You do not want to keep data around any longer than is necessary. Data stored in an `HttpServletRequest` object is lost after a response is given. In many cases, this is sufficient. However, if you were writing a shopping cart application, you would want to preserve whatever a client chooses as they move from page to page on the site. Here is where an `HttpSession` object is used.

Let us now look at a file that we can use to configure how a server deals with servlets, called `web.xml`.

Configuring deployment with the web.xml file

In the `WEB-INF` folder of a web project there is usually a file named `web.xml`. It was mandatory before the `@WebServlet` annotation was introduced. With the annotation, the application server can determine on its own which files are servlets and which are not. There is more that you can do in this file than just list servlets. For this reason, I advise you to always have a `web.xml` file.

Our descriptor will be quite basic:

```
<web-app xmlns="https://jakarta.ee/xml/ns/jakartaee"
    xmlns:xsi="http://www.w3.org/2001/XMLSchema-instance"
    xsi:schemaLocation=
              "https://jakarta.ee/xml/ns/jakartaee
       https://jakarta.ee/xml/ns/jakartaee/web-app_5_0.xsd"
       version="5.0' >
    <display-name>BasicServlet</display-name>
    <welcome-file-list>
        <welcome-file>index.html</welcome-file>
    </welcome-file-list>
</web-app>
```

We have a display name that the application server can use in a report, followed by the welcome page. The welcome page is the name of the page to display if the URL does not include the page name. Let's say you type the following in your browser:

```
http://localhost:8080/BasicServlet/index.html
```

Instead of writing that, you only need to write the following:

```
http://localhost:8080/BasicServlet
```

The HTTP protocol is stateless. This means that every time you make a request, the server behaves as if it is the first time you have visited the site. The application server can remember you by using an `HttpSession` object. This object has a default lifetime of 30 minutes since your last visit to the website. When the time is up, the object is invalidated, and the server will no longer remember you. You can change the length of time by adding this to the `web.xml` file:

```
<session-config>
    <session-timeout>
        30
    </session-timeout>
</session-config>
```

In some cases, you may have data, in the form of a string, which is common to every servlet in the application – for example, the company email address that needs to appear on every page. We use `context-param` for this:

```
<context-param>
    <param-name>email</param-name>
    <param-value>me@me.com</param-value>
</context-param>
```

To access this in a servlet, we just need the following:

```
String email =
        getServletContext().getInitParameter("email");
```

You should now be able to get a web application based on a servlet up and running.

Summary

In this chapter, we looked at the basics of a web application. The center of this universe is the servlet. There are many other frameworks, such as Spring, that provide an alternative set of libraries, yet all these frameworks sit on top of and depend upon the servlet specification, along with other Jakarta libraries.

Jakarta is standards-based. What this means is that by adhering to the HTTP protocols, it can provide services to any frontend, such as React.js, Bootstrap, and Angular. In the next chapter, we will look at one frontend programming library, Jakarta Faces, that is part of the Jakarta framework.

We used the GlassFish server in this chapter, but there are a number of other choices for a Java application server. For example, the Payara server is based on Glassfish, but as it is backed by the Payara company, it provides commercial support that is not available with Glassfish. There are also servers from Red Hat, IBM, and others. There is usually a community version that you can use without paying for a commercial license.

As we looked at server-side programming, we needed to make changes to our Maven pom.xml file. With these in place, we were able to create a .war file for use on the server as easily as we created desktop .jar files.

Next up, we will look deeper into Jakarta EE by examining an application that brings the Financial Calculator we saw in the previous chapter to the web.

Further reading

- Jakarta EE: https://jakarta.ee/
- HTTP methods: https://www.tutorialspoint.com/http/http_methods.htm

15

Jakarta Faces Application

Jakarta Faces, now just called Faces, is one of two client rendering techniques available in web applications. The other is **Jakarta Server Pages (JSP)**. In this chapter, we will examine a Faces web application that, like our Swing and JavaFX examples, allows you to perform three common finance calculations.

The JSP rendering approach permits the placement of Java source code on an HTML page. Before a JSP page can be rendered, the file is converted into a servlet by the application server. If you have 50 JSP pages, then there will be 50 servlets on the application server. The typical approach in designing an application is to use JSP for rendering by mixing standard HTML, expression language code to access data or call Java methods, and Java source code. These files end with a .jsp extension. While you can do processing on the page, the common approach is to have a JSP page call upon a servlet for processing and to decide which JSP page to return to the browser.

The Faces approach is quite different. First off, the framework provides the Faces servlet. All requests for a .jsf page are processed by this servlet. While a JSP application is usually a combination of .jsp pages and servlets, a Faces application does not require any servlets, though they can be used. In the place of servlets, Faces allows you to use a **plain old Java object (POJO)** for any processing the page requires. We will see that, with minor changes, we can use our Calculations and FinanceBean classes from *Chapter 13, Desktop Graphical User Interface Coding with Swing and JavaFX*, by adding some annotations to the code.

A Faces page contains Facelets along with any standard HTML tag. Facelets is the **view declaration language (VDL)**. As we will see shortly, a Faces page looks like an HTML page but uses custom tags called Facelets to describe how the page is rendered. This is significant because these Facelets are calling upon methods in the Faces framework that return standard HTML and JavaScript. This means that you can create your own custom tags and matching Java code. There are companies that provide libraries of Facelets that you can use by simply adding the library to your Maven pom file. See the *Further reading* section for examples of these third-party libraries.

We will interact with most objects by employing **Context Dependency Injection (CDI)**. When a class is annotated for CDI, you no longer need to instantiate the object with new. Instead, the CDI framework instantiates an object upon first use and determines whether it needs to be garbage collected or not. We can also use **Bean Validation** (**BV**) in a CDI bean. This allows you, using annotations, to define how to determine whether an assignment of data through a setter method is valid. No need for an `if` statement.

This chapter will now take you through how our Financial Calculator is written using Faces, CDI, and BV. We will examine the following topics:

- Configuring a Faces application
- Creating an object managed by CDI and validated with BV
- Using XHTML, Facelets, and Jakarta Expression Language for rendering pages
- Understanding the life cycle of a Faces page

By the end of this chapter, you will understand how a Faces web application is constructed. With this knowledge, you will be able to evaluate other web application frameworks, such as Spring and Vaadin.

Technical requirements

For this chapter, you'll need the following:

- Java 17
- A text editor
- Maven 3.8.6 or a newer version installed
- GlassFish 7.0 application server
- Web browser

Sample code for this chapter is available at `https://github.com/PacktPublishing/Transitioning-to-Java/tree/chapter15`.

Configuring a Faces application

Here is what the web version of the Financial Calculator app looks like:

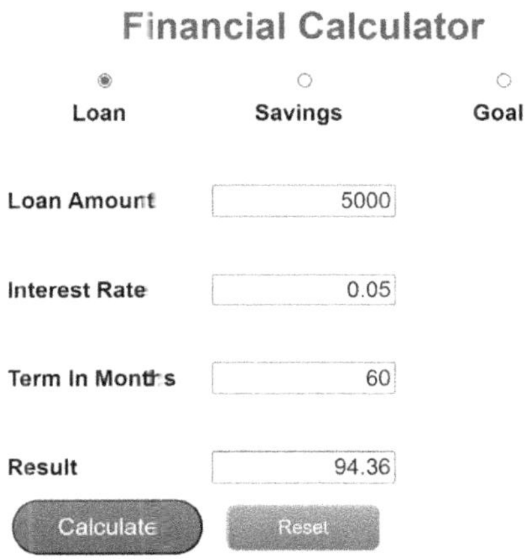

Figure 15.1 – The Financial Calculator web page

Configuring a Faces project begins with the same setup as any basic web application, such as what we saw in the previous chapter. The Maven folder setup is identical. In the WEB-INF folder, we have three required XML configuration files and one that is optional. Let us begin with beans.xml:

```xml
<beans xmlns="https://jakarta.ee/xml/ns/jakartaee"
    xmlns:xsi=http://www.w3.org/2001/XMLSchema-instance
    xsi:schemaLocation="https://jakarta.ee/xml/ns/jakartaee
    https://jakarta.ee/xml/ns/jakartaee/beans_3_0.xsd"
        bean-discovery-mode="annotated">
</beans>
```

This looks strange because there is only one tag, beans. Before the widespread usage of annotations, listing every bean or class was necessary to enable CDI. The bean-discovery-mode tag defines any bean with a CDI annotation is now subject to the CDI. Prior to Jakarta EE 10, the discovery mode to use was all. The current best practice is to use annotated rather than all.

The next configuration file is faces-config.xml. In this file, you can define application properties. Some of these properties can be navigation rules to determine to which page should a submit request go next, bean objects that should be instantiated, and message bundles for i18n support. In this example, we are only using this file to define the message bundle for this application:

```xml
<faces-config version="4.0"
    xmlns="https://jakarta.ee/xml/ns/jakartaee"
    xmlns:xsi="http://www.w3.org/2001/XMLSchema-instance"
    xsi:schemaLocation=
```

```
                  "https://jakarta.ee/xml/ns/jakartaee
                  https://jakarta.ee/xml/ns/jakartaee/
                                  web-facesconfig_4_0.xsd">
    <application>
        <resource-bundle>
            <base-name>
                com.kenfogel.bundles.MessagesBundle
            </base-name>
            <var>msgs</var>
        </resource-bundle>
    </application>
</faces-config>
```

You can see in base-name the package and base file name for the message bundles. Under var you can see the name of the identifier we can use on a Faces XHTML page. Bundles in Faces are identical to how we created resource bundles that are configured for desktop applications with a key followed by a value. To refer to a value from a message bundle, we use expression language such as #{msgs. result}.

The last required configuration file is the web.xml file. It can fulfill the same responsibilities as we saw in the previous chapter. In addition, we can modify how Faces performs. For simplicity, I have removed the opening and closing tags as they are identical to the previous chapter's version.

The first param, PROJECT_STAGE, configures the framework error handling. Using the Development error messages carries more information than a small code of slower performance. Typically, you will change this to Production when the code is complete:

```
<context-param>
    <param-name>jakarta.faces.PROJECT_STAGE</param-name>
    <param-value>Development</param-value>
</context-param>
```

The next param determines whether comments in the Faces page will be in the HTML sent to the browser or not:

```
<context-param>
    <param-name>
        jakarta.faces.FACELETS_SKIP_COMMENTS
    </param-name>
    <param-value>true</param-value>
</context-param>
```

The server will destroy a session object and send it for garbage collection if it is explicitly destroyed or the session timeout is reached:

```
<session-config>
    <session-timeout>
        30
    </session-timeout>
</session-config>
```

When referring to a website by its name rather than a specific page, then this is the page that will be displayed. In Faces, it should always be a .xhtml file:

```
<welcome-file-list>
    <welcome-file>index.xhtml</welcome-file>
</welcome-file-list>
```

The last and optional configuration file is glassfish-web.xml. We can provide configuration information for database connection pooling, security information, and other items that the application server is responsible for in this file. In my example projects, I have removed this file as we do not require it.

With a project organized and configuration files in place, we are ready to start on our application. Before we concern ourselves with the web page design, we need to set up the POJOs that our application will require and configure them to work under the care of CDI.

Creating an object managed by Context Dependency Injection and validated with Bean Validation

Only two Java classes are used in this program, and they are nearly identical to what we used in *Chapter 13, Desktop Graphical User Interface Coding with Swing and JavaFX*. They are both subject to CDI, and the data class also uses BV. Rather than show the entire code for these beans we saw in *Chapter 13*, we will only look at what needs to be changed.

FinanceBean

The first annotation, @Named, defines this class as under the control of CDI. Before CDI was widely used, JSF had its own CDI-like implementation that used the @ManagedBean annotation. This is considered obsolete and should no longer be used. The name in the parenthesis, money, is an alias we can use in the expression language:

```
@Named("money")
```

Scopes

When an object managed by CDI in a Jakarta application is created or destroyed, and other classes may access it, it is referred to as the Scope. There are the following types:

- @RequestScoped: This means that the server will create a new object for every request, and the previous object from a previous request is sent out for garbage collection for each user.

- @SessionScoped: This means that objects created upon the first request remain in place and are only destroyed explicitly or when the session timer ends for each user.

- @ApplicationScoped: These objects are available to every session for all users.

- @ViewScoped: This is the final scope. Beans created with this scope are tied to a Faces page. As long as you do not change the view, such as by having a link or button that calls upon another page, then the bean remains valid.

Now back to the code:

```
@RequestScoped
public class FinanceBean implements Serializable {
    private static final Logger LOG =
        Logger.getLogger(FinanceBean.class.getName());
```

Each of the BigDecimal variables has been declared with BV annotations. In this example, we are setting a minimum and maximum value. The message attribute is the key to a separate message bundle that is named ValidationMessages. Just like ordinary bundles, you need a default and then one for each supported language. These validation bundles are expected to be found in the resources folder and not in any folders below it:

```
@DecimalMin(value = "1.00",
        message = "{com.kenfogel.minInput}")
@DecimalMax(value = "100000.00",
        message = "{com.kenfogel.maxInput}")
private BigDecimal inputValue;
@DecimalMin(value = "0.00",
        message = "{com.kenfogel.minInput}")
@DecimalMax(value = "1.00",
        message = "{com.kenfogel.maxInput}")
private BigDecimal rate;
@DecimalMin(value = "1.00",
        message = "{com.kenfogel.minInput}")
@DecimalMax(value = "300.00",
```

```
        message = "{con.kenfogel.maxInput}")
    private BigDecimal term;
    private BigDecimal result;
```

Here are two new fields not found in the original `FinanceBean` class. The first is the `calculationType` field that defines which of the three calculation formulas are used. It is also used to update the name of `Label` of the first input field:

```
    private String calculationType;
```

The new text must be shown in the first input label when the calculation type changes. This will be read from the resource bundle:

```
    private final ResourceBundle msgs;
    public FinanceBean() {
        result = BigDecimal.ZERO;
        inputValue = BigDecimal.ZERO;
        rate = BigDecimal.ZERO;
        term = BigDecimal.ZERO;
```

Here in the constructor, we define the calculation as `loan` and initialize `msgs`:

```
        calculationType = "loan";
        msgs = ResourceBundle.getBundle(
            "com.kenfogel.bundles.MessagesBundle");
    }
```

The remaining methods in this data class are just the usual getters and setters.

One last point about CDI and BV is that they can be used in any type of Java application that includes the CDI and/or BV library. That library is part of Jakarta, so there is no specific reference to it in the pom file. To use just CDI, BV, or both in your application, add the following to your pom file:

```
<dependency>
    <groupId>jakarta.enterprise</groupId>
    <artifactId>jakarta.enterprise.cdi-api</artifactId>
    <version>4.0.1</version>
</dependency>
<dependency>
    <groupId>jakarta.validation</groupId>
    <artifactId>jakarta.validation-api</artifactId>
```

```
        <version>3.0.2</version>
    </dependency>
```

With a data bean called a backing bean in Faces, in place, we can now look at the `Calculations` class.

Calculations

The `Calculations` class is also mostly unchanged. The formulas in all three calculation methods are the same. The first change is that the `FinanceBean` object is now a class field that is instantiated by CDI rather than a parameter passed to each method. The second change is that calls for a calculation are made to a method that, in turn, selects the appropriate calculation method. Let us look at this now.

The class begins with the annotation that defines this as a CDI-managed bean. The scope is @ `RequestScope`. A CDI bean is instantiated either when it is injected into another class, as we will see next, or upon first use on a Faces page:

```
@Named("calculate")
@RequestScoped
public class Calculations implements Serializable {
    private static final Logger LOG =
        Logger.getLogger(Calculations.class.getName());
```

With the `@Inject` annotation, CDI will check whether this object currently exists. If it does, then a reference to it is assigned to a variable named `money`. If it does not exist, then it will be instantiated before passing the reference into `money`:

```
@Inject
FinanceBean money;
```

This method will be called from the Faces page:

```
public String performCalculation() {
    switch (money.getCalculationType()) {
        case "loan" ->
            loanCalculation();
        case "savings" ->
            futureValueCalculation();
        case "goal" ->
            savingsGoalCalculation();
    }
    return null;
}
```

This is the start of one of the calculations that uses the class field to get the user input and where the result will be stored:

```
public void loanCalculation()
        throws ArithmeticException {

    // Divide APR by 12
    var monthlyRate = money.getRate().divide(
        new BigDecimal("12"), MathContext.DECIMAL64);
```

We have only scratched the surface of CDI and BV. See *Further reading* to find additional information on these features. Now let's move on to Faces rendering of web pages.

Using XHTML, Facelets, and Expression Language for rendering pages

Faces applications use files with an xhtml extension. This extension means that either HTML or custom tags, called Facelets, must adhere to the rules of XML. This means that every tag must be closed. HTML allows for tags such as <br> and <p>, while to use these tags in XHTML, there must be an opening tag followed by a closing tag. Tags can also be self-closing by ending in a forward slash such as <br/> or <p/>.

Let us look at the index.xhtml file that is responsible for the user interface.

We begin by declaring that this file is in the XHTML format:

```
<!DOCTYPE xhtml>
```

XML documents are checked to ensure all tags are valid. The five namespaces listed here represent the common set of tags available in Faces:

```
<html xmlns:faces="jakarta.faces"
      xmlns:ui="jakarta.faces.facelets"
      xmlns:f="jakarta.faces.core"
      xmlns:h="jakarta.faces.html"
      xmlns:pt="jakarta.faces.passthrough" >
```

Here we see our first Facelet, the h:head tag. When this file is processed by the Faces framework, each Facelet is a call to a Java method that returns a valid HTML string and, if needed, JavaScript:

```
<h:head>
```

Here we see our first Expression Language statement. In this case, we are retrieving the text from the messages bundle defined in the `faces-config.xml` file. Notice also that we are using an HTML tag and a title, and these tags are preserved in the HTML generated from processing the Faces page. In any situation where a Facelet matches an HTML tag, you should always use the Facelet:

```
<title>#{msgs.title}</title>
<h:outputStylesheet library="css" name="main.css"/>
</h:head>
<h:body>
    <h:form>
        <h1>#{msgs.heading}</h1>
```

Here we can see the Facelets for radio button input. When we call upon a method in expression language without ending the call with parenthesis, we are indicating that we want either `getCalculationType()` or `setCalculationType()`. We must use a method's full name followed by parenthesis if it is not a setter or getter:

```
<h:selectOneRadio
    value="#{money.calculationType}"
    immediate="true" styleClass="radiocenter" >
    <f:selectItem itemValue="loan"
        itemLabel="#{msgs.loan_radio}"
        styleclass="radiocenter"/>
    <f:selectItem itemValue="savings"
        itemLabel="#{msgs.savings_radio}"
        styleclass="radiocenter"/>
    <f:selectItem itemValue="goal"
        itemLabel="#{msgs.goal_radio}"
        styleclass="radiocenter" />
```

The common use of radio buttons is to provide a choice required when the form is submitted. In the design of this application, I wanted the fields to be cleared and the form rendered again. This re-rendering will also change the text of the input label of the first label. The `valueChange` event indicates that an Ajax partial submit will occur that will call upon the `money.clear()` method to reset all values to zero. The `render="@form"` attribute will result in the page being re-rendered:

```
<f:ajax event="valueChange" render="@form"
    action="#{money.clear()}"/>
</h:selectOneRadio>
```

Here we are using `panelGrid`, which creates an HTML table. You indicate the number of columns, while the number of rows is determined by the number of either HTML tags or Facelets. The first value every two rows is a non-breaking space. This will consume a cell in the table but display nothing:

```
<h:panelGrid columns="2" >
    <h:outputLabel value=" "/>
```

The second value is `h:message`. This Facelet defaults to a blank entry. If an error occurs such as invalid input or a value out of range, then a message will appear above the input field. You can use either a `style` attribute to write the CSS in this attribute or use `styleclass` to refer to a class in the CSS file:

```
<h:message id = "inputError"
    for="inputValue"
    style="white-space:nowrap; color:red;
    font-size: 100%;"/>
```

Here is what will appear if the user enters invalid or unconverted input. The appearance of these messages, along with everything else on a Faces page, can be styled with CSS:

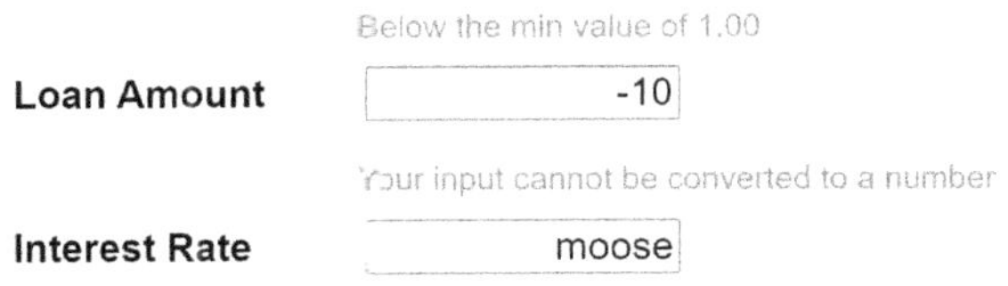

Below the min value of 1.00

Loan Amount -10

Your input cannot be converted to a number

Interest Rate moose

Figure 15.2 – The h:message output

The text for this input label is retrieved from `FinanceBean` rather than from a message bundle directly. This is how the label can change based on the radio button choice.

Each `h:inputText` field contains a `f:ajax` Facelet. This will issue a partial submit, allowing the string you entered to be converted to `BigDecimal` and then checked whether it is in range. Otherwise, these checks will only occur when a **Submit** button is pressed. There is nothing more annoying than filling in a form only to discover several input errors after the **Submit** button is pressed.

The Faces framework takes care of the conversion from `String` to `BigDecimal`. If this fails due to the presence of invalid characters, the matching `h:message` field will appear with a message from the message bundle file. The `converterMessage` attribute contains the key value for the bundle:

```
<h:outputLabel id = "inputLabel"
    value="#{money.getInputLabel()}"
    for="inputValue" />
```

```
<h:inputText value="#{money.inputValue}"
  id="inputValue"
  converterMessage="#{msgs.invalidInput}" >
  <f:ajax event="blur"
    render="inputError" />
</h:inputText>
<h:outputLabel value=" "/>
<h:message id="interestError"
    for="interestValue"
    style="white-space:nowrap;
    color:red; font-size: 100%; " />
<h:outputLabel value="#{msgs.interest}"
    for="interestValue"/>
<h:inputText value="#{money.rate}"
  id="interestValue"
  converterMessage="#{msgs.invalidInput}" >
    <f:ajax event="blur"
        render="interestError" />
</h:inputText>
```

There are two more rows on this form that I have removed in the text as they are nearly identical to the previous one.

At the bottom of our form are two buttons. One invokes the Calculations class to generate the answer, while the second resets all the fields and makes Load the choice in the radio button:

```
<h:commandButton type="submit"
  action="#{calculate.performCalculation()}"
  value="#{msgs.submit}"  styleClass="btn" />
<h:commandButton type="reset"
  value="#{msgs.clear}" styleClass="btn2" >
    <f:ajax event="click" execute="@this"
    render="@form" />
</h:commandButton>
    </h:panelGrid>
  </h:form>
```

This is a basic application, but it should give you a sense of how Faces applications work.

Deploying a Faces web application

Like every sample program in this book that is built with Maven, all you need to do is open a terminal/console window in the `project` folder. At the prompt, you just need to enter mvn. Assuming that there are no errors, you should find your project in the target folder.

You can copy this file and paste it into the `autodeploy` folder discussed in the previous chapter. The other option is to deploy the application from the GlassFish console:

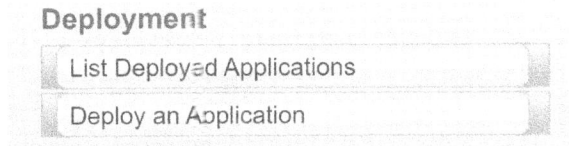

Deployment

| List Deployed Applications |
| Deploy an Application |

Figure 15.3 – Deploying from the server

Selecting **Deploy an Application** will bring you to a form where you can upload your application to the server. With our application up and running, let us take a deeper look into what happens when we interact with a Faces page.

Understanding the life cycle of a Faces page

In *Chapter 14, Server-Side Coding with Jakarta*, we saw the basic life cycle of web apps that employ servlets. Simply put, a submit request is made to a servlet, the servlet receives data from the page in the form of a request object, and you code whatever tasks are necessary, and then a response is returned either from the servlet or as an HTML or JSP page. Faces works differently from this.

There are six parts to the life cycle of a Faces page that begins with a request for a `.jsf` page. Here is a diagram that shows the steps in the Faces life cycle:

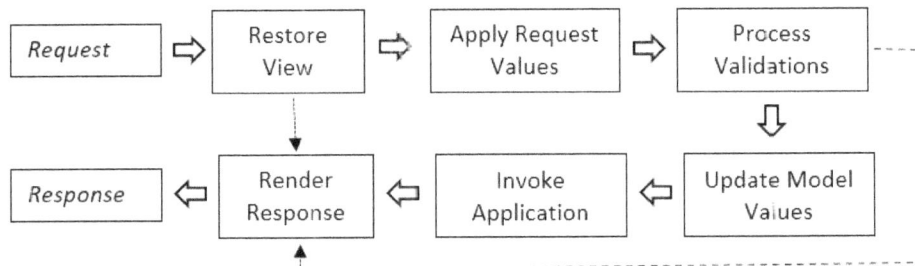

Figure 15.4 – The Faces life cycle

Let's review each part:

- **Restore View**: When a request arrives, it is checked for a query string. If it is not present, then this is likely the first time this page is requested. This means that the page can be rendered without the need to go through any of the other phases.

- **Apply Request Values**: In this phase, the contents of the query string are assigned `String` variables. This allows for validating and converting the data before it is assigned to the bean associated with this request. Beans that are associated with a page are called backing beans.

- **Process Validations**: In this phase, the data in each field in the query string is first converted to the matching type in the backing bean. As all input is in string form, it must be converted to the type found in the backing bean. There are standard converters, such as turning `String` to `BigDecimal` or `double`. You can also write your own custom converters. If anything goes wrong in the conversion, the remaining phases are ignored, and the **Render Response** phase is invoked. This is where `h:message` is invoked to present an error message.

  After the conversion comes validation. There are standard validator facelets as well as letting you write a custom validator. BV, if used, is also invoked here. If the validation fails, then, like a failed conversion, the life cycle jumps to Render Response.

- **Update Model Values**: In this phase, after a successful conversion and validation, the values are assigned to the backing bean.

- **Invoke Application**: Many tags have an action attribute that allows you to call upon methods in a backing bean. With the data now in the backing bean, these methods can be called.

- **Render Response**: Here, the Faces page is rendered as HTML and JavaScript.

It is possible to write a phase listener class that you can use to add additional tasks in most phases. Understanding the life cycle is critical to developing Faces pages.

Summary

This chapter was a brief introduction to Jakarta Faces, the supporting frameworks such as CDI and BV, and how to deploy an application. Looking at the life cycle should give you an understanding of what the Faces servlet is doing. While web page rendering is most commonly the domain of JavaScript frameworks, Faces should be considered as an alternative to the JavaScript approach. Its integration with CDI and BV makes it a solid foundation for web apps. BV can ensure that all validation is done on the server. This does not preclude using validation in JS. However, with a simple tool such as `curl`, you could easily submit invalid data if validation only occurred on the client side in JS.

In our next and final chapter, we will look at how Java applications can be packaged for easy deployment to a server or distributed for easy use as a desktop application.

Further reading

- PrimeFaces – Facelet library: `https://www.primefaces.org/`

- OmniFaces – Faces utility methods: `https://omnifaces.org/`

- Spring: `https://spring.ic/`

- Vaadin: `https://vaadin.com/`

- Jakarta Tutorial: `https://eclipse-ee4j.github.io/jakartaee-tutorial/`

Part 4:
Packaging Java Code

This part is about how you distribute your code. You will learn how to create a custom Java runtime and package it with an application in a single file installer. The distribution of an entire environment, including Java, an application server, and your app using Docker, the first step to cloud deployment, is the final topic.

This part contains the following chapter:

- *Chapter 16, Deploying Java in Standalone Packages and Containers*

16
Deploying Java in Standalone Packages and Containers

In this last chapter, we will look at different ways to package and distribute a Java application. We have already seen JAR files for desktop applications and WAR files for web applications, along with how to deploy them. While this approach can be sufficient for deployment, there are situations where this traditional approach can be improved upon.

Java is big. There are numerous libraries in the Java SE distribution, although it is likely that your application only needs some of them. The same can be said for third-party or external libraries. Modern packaging using the Java module approach allows you to produce JAR or WAR files that only contain parts of a library that you will use.

In the case of web applications, this type of packaging can reduce the size of a WAR file to contain only the required modules from a required external library, rather than the entire library. In the case of desktop applications, it is required that the Java language must already be installed on a computer. The Java runtime is now modularized. This allows you to create executable applications that do not require an installed version of Java to run but, rather, include it as part of the packaged installer.

Finally, we will look at the Docker container system. Imagine a team of developers, each with different **operating systems** (**OSes**), working on an application. While Java is *write once, run anywhere*, it is sometimes advantageous to have every developer working in an identical environment. Docker containers help meet this need. Furthermore, you can deploy these containers to a cloud. While we will not look at cloud deployment, understanding how containers work prepares you for working within the cloud.

We will cover the following in this chapter:

- Exploring what modular Java is
- Creating a custom JRE with `jlink`
- Packaging with an installer using `jpackage`

- Using the Docker container system
- Working with Docker images
- Creating a Docker image
- Publishing an image

We will use a modified version of BankSwing, originally from *Chapter 13, Desktop Graphical User Interface Coding with Swing and JavaFX*, and now renamed BankSwingJUL in this chapter, to explore modules and packages. To look at Docker, we will use JSF_FinancialCalculator, unchanged from the previous chapter.

Technical requirements

You'll need the following for this chapter:

- Java 17.
- A text editor.
- Maven 3.8.6 or a newer version.
- For jpackage:

  - **Windows**: The WiX toolset (https://wixtoolset.org/)
  - **Red Hat Linux**: The rpm-build package
  - **Ubuntu Linux**: The fakeroot package
  - **macOS**: Xcode command-line tools

- Docker Desktop (https://www.docker.com/). To use Docker, you will need to create an account. The free Personal account is sufficient. Once you have an account, you can download Docker Desktop. There is a version for every OS.

The sample code for this chapter is available at https://github.com/PacktPublishing/Transitioning-to-Java/tree/chapter16.

Exploring what modular Java is

Up to this point, we saw code in classes that consists of class fields and methods. Then, we grouped these classes into packages and, finally, as a JAR or WAR file. Modular Java introduces a new grouping called **modules**. A module is a JAR file but with a module descriptor. There is also an automatic module that has a module name in its manifest file. This feature of Java is called the **Java Platform Module System (JPMS)**.

Up until now, we used Maven to build our applications. Maven is a build tool that downloads any libraries we need and ensures that our code will compile successfully. What it does not do is determine whether all the required external libraries, such as Java itself, are present. Its primary job ends when the code successfully compiles. The JPMS, on the other hand, focuses on the libraries required to successfully run a program. Unlike Maven, JPMS checks that libraries coded as modules are present or will be present when the code runs. This leads to the question, what is a module?

A module is a JAR file. There are minor differences between a regular JAR file and a module JAR file. At a minimum, a module file must have a file named `module-info.java` in the `src/main/java` folder. One purpose of this file is to list the required modules. There may not be any required modules, but the presence of this file denotes that this project can be a module. Not every library coded in Java has been recoded to be a module, but many new libraries are coded this way. A module file can be used as an ordinary JAR file or as a module when using JPMS tools. You do not need two versions of a library.

At one time, there were two versions of Java available to users. There is the JDK that contains the JVM and all the required developer tools, such as the Java compiler. The second version was the **Java Runtime Edition** (**JRE**). As its name implies, the JRE contains all the necessary libraries to run almost any Java program. The JRE was significantly smaller, coming in at around 90 MB, while the full JDK is around 300 MB. With the introduction of the JPMS, the JRE was no longer available as a download. Times change, and some Java distributions now contain a JRE again.

With the JRE significantly smaller than the JDK, what can modules do for us? The reason is related to why the JRE was dropped from Java distributions. With the JPMS, you can construct your own custom JRE, including only those modules you need. So, what are the modules in the Java language? In a terminal/console window, enter the following:

```
java --list-modules
```

You will now get a list of every module. On my Windows 11 system, there are 71 modules listed – 22 that begin with `java` and 49 that begin with `jdk`. To build a custom JRE, you need to know which modules your program uses. Retrieve from this chapter's GitHub the `BankSwingJUL` project. The only difference from the *Chapter 13* version is that JUL replaces `log4j2`. I have done this to reduce the number of modules required to just those in the Java distribution. Build the project, and you should find in the `target` folder a JAR file named `BankSwingJUL-0.1-SNAPSHOT.jar`. Open a terminal/console window in the `target` folder and enter the following:

```
jdeps BankSwingJUL-0.1-SNAPSHOT.jar
```

The output will begin with a summary of the Java modules you will need:

```
BankSwingJUL-0.1-SNAPSHOT.jar -> java.base
BankSwingJUL-0.1-SNAPSHOT.jar -> java.desktop
BankSwingJUL-0.1-SNAPSHOT.jar -> java.logging
```

The remainder of the output looks at every class in the project, showing you what Java classes you are using and what module they belong to. The `java.base` module is the home to the core set of classes. The `java.desktop` module is the home of Swing, while `java.logging` module is the home of JUL. Now, it's time to create our custom JRE.

Creating a custom JRE with jlink

We will use the `jlink` tool that is part of the Java JDK to create our custom JRE. We will begin by creating a JRE that contains all the required modules:

```
jlink --add-modules ALL-MODULE-PATH --output jdk-17.0.2-jre
        --strip-debug --no-man-pages --no-header-files
        --compress=2
```

This is one line. In Linux, you can enter a multiline command using the backslash (\), while in Windows, you use the caret (^). The output of this command will be a folder named `jdk-17.0.2-jre` that contains a JRE of only 76 MB. This is smaller than the original JRE, but we do not want all the Java modules; we just need three. Here is our new command:

```
jlink --add-modules java.base,java.desktop,java.logging
        --output jdk-17.0.2-minimaljre --strip-debug
        --no-man-pages --no-header-files --compress=2
```

We will now have a new JRE in the `jdk-17.0.2-minimaljre` folder that is only 41 MB. Now, we need to use our custom JRE with our application. To test whether our JRE is working, you can execute the application by first opening a terminal/console window in the `bin` folder of the custom JRE you have created. Issue the following command to run your code. Take note that the paths are for Windows, so they must be adjusted for Linux or Mac:

```
java.exe -jar C:\dev\Packt\16\BankSwingJUL\target\
    BankSwingJUL-0.1-SNAPSHOT.jar
```

This is a single-line command. If all is well, your `BankSwingJUL` app will run. Now, it's time to wrap up the application into a single executable file that contains both our application and the JRE. This will allow us to distribute our applications without requiring the recipient of our program to first install Java.

Packaging with an installer using jpackage

With our custom JRE created, we are now ready to create a custom installable package. You can create these for Windows, Linux, or Mac. You must use the OS that is the target of your package. In addition, there are additional steps for each OS.

Windows requires you to install the WiX toolset. You can find this at `https://wixtoolset.org/`. Download the latest version and install it. When you run `jpackage`, it will produce an EXE file. You can distribute this file, and when run, it will install all that is necessary to run the program in the `C:\Program Files` directory. An executable EXE file will be in the folder, and this is how you will run your program.

Linux users, depending on the version they are using, will need the `rpm-build` or `fakeroot` package. When you run `jpackage`, it will produce a DEB file for Debian Linux or an RPM file for other distros. You can distribute this file, and when run, it will install all that is necessary to run the program in the `/opt/application-name` directory. An executable file will be in the folder, and this is how you will run your program.

Mac users require the Xcode command-line tools. When you run `jpackage`, it will produce a DMG file. You can distribute this file, and when run, it will install all that is necessary to run the program in the `/Applications/application-name` directory. An executable file will be in the folder, and this is how you will run your program.

In all three cases, it is not necessary to have Java installed. Even if you do, you will be using the custom JRE.

To create an installer package with `jpackage`, you simply enter the following on the command line:

```
jpackage --name BankSwingJUL
   --input C:\dev\Packt\16\BankSwingJUL\target
   --main-jar BankSwingJUL-0.1-SNAPSHOT.jar
   --runtime-image C:\dev\Packt\16\jre\jdk-17.0.2-minimaljre
   --dest C:\temp
```

This is a single-line command. Here is a rundown of the parameters:

- `--name`: The name of the executable file with `-1.0` added. Use `-app-version` followed by a version designation to override this.

- `--input`: The location of the JAR file you are packaging.

- `--main-jar`: The name of the JAR file that contains the class with the `main` method. If you do not have a `MANIFEST.MF` file in your JAR file that lists the class with a `main` method, you can use `-main-class`, followed by the name of the class that contains the `main` method.

- `--runtime-image`: This is the path and name of the JRE folder you created with `jlink`.

- `--dest`: By default, the packaged application will be found in whatever folder you issued the `jpackage` command. You can choose the folder you want with this parameter.

Upon the successful conclusion of this command, you will have an executable package that will install your program, with an executable file to run it.

Web applications depend on an application server and not the JRE to run. For this reason, we cannot use jpackage. This is where our next choice for packaging comes in, the Docker container.

Using the Docker container system

Docker is a platform-as-a-service system that allows you to construct an image of a running application that can run in a virtualized Linux container. These images can be run on any computer that supports Docker containers. This includes Windows and Linux distributions. This image can contain everything necessary to run the program. In this section, we will create an image with a Java application server, a Java JDK, and our JSF_FinancialCalculator web application and deploy it in a container. Why this is significant is that most cloud providers, such as AWS, support the deployment of cloud applications in Docker containers. We will not be discussing cloud deployment, as the various cloud providers work differently. What they share is the use of Docker.

The first step is to install the Docker system. The easiest way is to download and install the Docker Desktop system from https://www.docker.com/. There is a version each for Windows, Mac, and Linux, and they contain a GUI interface as well as command-line tools. On a Windows 10 or 11 system that supports WSL2, the command-line tools are available in both, a Windows terminal and a WSL2 Linux Terminal. This means that, except for how paths to files are described, all commands work the same on all OSes. Now, take a moment and install Docker.

Working with Docker images

While we could build an image from scratch, there is another way. Many organizations that create software that is destined for the cloud make available pre-built images. To these images, we can add our application. In our case, we want a pre-built image with Java 17 and an application server. We will use an image from Payara. This company provides a server based on GlassFish, with enhancements in both an open source community version and a commercial paid version.

Images on Docker Hub may have been created for malicious reasons. While Docker provides a service to scan for vulnerabilities, you should also scan any executable files in an image for potentially malicious behavior. The Docker plan you have signed up for determines how many images you can pull from or push to the Hub. With the free Personal subscription, you may have an unlimited number of public repositories you can push, but you are restricted to no more than 400 image pulls a day. The commercial subscriptions increase the number of pulls from the repository and can carry out vulnerability scans on your image.

Start Docker Desktop. It comes with an image and container that contains a basic web server that has the documentation pages for Docker. We will do most of our setup on the command line, while the desktop GUI is useful for seeing what the state of Docker images and containers are.

The first step is to download the image we will modify by adding the JSF_FinancialCalculator application. We will use this program unchanged from the previous chapter. Here is the command:

```
docker pull payara/server-full:6.2023.2-jdk17
```

If you visit `https://hub.docker.com/r/payara/server-full/tags`, you can see all the versions of the Payara server available. As you can see from the previous command, we are pulling the `server-full:6.2023.2-jdk17` image that contains both the server and Java 17. In Docker, a successful command returns a long stream of digits.

Now, we need to run this image in a container. While you can run multiple containers, network applications that use TCP ports can result in conflicts. For this reason, I recommend stopping any containers that are currently running. Using the Docker Desktop, select the container list from the menu and look for any containers listed as **Running**, and then stop them by clicking on the square button in the **Actions** column. You can also stop a container by entering the following at the command line:

```
docker stop my_container
```

Here, `my_container` is replaced by the name of a running container or image.

We now want to wrap this image in a container and run the image:

```
docker run --name finance -p 8080:8080 -p 4848:4848
            payara/server-full:6.2023.2-jdk17
```

This is a single-line command. The `--name` switch allows you to assign a name to the container. If you leave this switch out, Docker will assign a random name. The `-p` switch maps a port in the container to a port of the computer. In this example, we are mapping to the same port. The name of the image is the same as the name of the image we pulled down. Assuming that there were no errors, you can now test the container. Go to your browser and first visit the Payara home page by entering `http://localhost:8080`. Next, visit the admin console page at `https://localhost:4848`. You may get a warning from your browser, as the TLS certificate is self-signed. Ignore the warning and you should get to the sign-in page. The username and password are both `admin`.

Under **Payara Server Console – Common Tasks**, look for the **Deployment** section and select **Deploy an Application**. Other than some changes in the background colors, the admin console is practically identical to GlassFish. When you select **Deploy**, you can select **Choose File** under **Packaged File to Be Uploaded to the Server**. You want to upload the WAR file from the `JSF_FinancialCalculator` example in the previous chapter, which you can find in the project's `target` folder.

You can now verify that the application has been properly deployed by entering in your browser `http://localhost:8080/JSF_FinancialCalculator`. The name of the project must match the name of the WAR file. If all works and the calculator opens in your browser, you can now create your own container based on the `payara/server-full:6.2023.2-jdk17` image, which will contain the calculator app installed on the server.

Creating a Docker image

Now, we are ready to create our own image. First, we need to stop the container we just used:

```
docker stop finance
```

In a terminal/console, enter the following command to create your new container, which will contain the Payara image:

```
docker create --name transition -p 8080:8080
        -p 4848:4848 payara/server-full:6.2023.2-jdk17
```

The name `transition` is arbitrary and can be anything you want. You now have a new container based on the Payara image. We want to modify this container to include the calculator application. The first step is to run this new container:

```
docker start transition
```

The most common error that occurs here is if another container is listening to the same ports. Ensure that any containers or images with Payara are not running. The Docker Desktop app can show you which containers or images are running.

Just as we did when we tested the Payara image, use your browser to open the admin console of Payara. Now, deploy the `JSF_FinancialCalculator` WAR file to the server. Verify that it is running successfully by visiting the application's web page.

Now, make the change to the image in the container, the addition of the web app, permanent by entering the following:

```
docker commit transition
```

There is one last step. Enter the following:

```
docker images
```

You will see an entry with `<none>` for both `REPOSITORY` and `TAG`:

```
REPOSITORY    TAG       IMAGE ID       CREATED          SIZE
<none>        <none>    c0236a80bba3   52 minutes ago   618MB
```

To resolve this, and as the final step in creating an image, assign a tag name and image ID by entering the following:

```
docker tag c0236a80bba3  transition-image
```

Take note that the hexadecimal number that must be used can be found in the table from the previous `docker images` command. When you run `docker images` after `docker tag`, the table will show the following:

```
REPOSITORY        TAG      IMAGE ID    CREATED       SIZE
transition-image  latest   c0236a80bba 32 hours ago 618MB
```

You now have a configured image in your local repository. For anyone to use your image, you must publish it on the Docker Hub website.

Publishing an image

As already noted, for security reasons, any image you use as the basis of a new image must be scanned for vulnerabilities, especially any executable code in the image. The free Personal tier allows you to have an unlimited number of public images. The paid tiers support private images. The first step in publishing is to create a repository on the Hub. To do this, open your browser and go to `https://hub.docker.com/`. Sign into your account if needed.

Next, select **Repositories** from the choices at the top of the web page. You will now see any repositories you may have already created. Click on **Create repository**. On this page, you must fill in the form, entering a name for the container along with an optional description. It also shows your Docker username. Ensure that **Public** is the choice for the repo type.

Now, you can push your image to the Hub. There are three steps:

1. Log in to Docker Hub:

   ```
   docker login --username my_username
   ```

   Replace `my_username` with your Docker username. You will now be asked for your password. You will receive confirmation of a successful login.

2. Next, you need to change the tag for your image, `transition-image`, to match the name of the repository you created, `omniprof/transitioning_to_java`. The name consists of your username and the name of the repository:

   ```
   docker tag transition-image omniprof/
      transitioning_to_java
   ```

3. Now comes the final step, pushing your image into the Hub:

   ```
   docker push omniprof/transitioning_to_java
   ```

To determine whether you were successful, visit Docker Hub and select **Repositories**. This time, you will see your repository named `omniprof/transitioning_to_java`.

You now have a Docker image that can be shared with your team or clients.

Summary

In this chapter, we looked at what modular Java means. We took advantage of the fact that Java itself has been modularized. This allows you to construct a JRE with `jlink` that is significantly smaller than the JDK. You can even make it smaller by only including the modules your code depends on.

We then looked at two ways to distribute your code. The first employed `jpackage` to create an installer for your application. The installer can include your custom JRE and will install your program, along with an executable file to run the application. This is usually the best way to distribute desktop applications.

The second distribution method uses the Docker container system. Docker allows us to construct and publish an image that includes not only our code and a JDK but also any other programs required. In our example, the extra program was an application server to which the finance application was installed. The images we construct are published to a repository, such as Docker Hub. Anyone running Docker on any OS can now pull our image and have it run in a Docker container.

This also leads us to the end of this book. My goal was to provide a reference to experienced developers in need of learning about and understanding Java. There is still much to learn, but my hope is that this book has put you on the right path.

Further reading

- *Multi-Module Maven Application with Java Modules*: `https://www.baeldung.com/maven-multi-module-project-java-jpms`

- *Java Platform, Standard Edition – Packaging Tool User's Guide*: `https://docs.oracle.com/en/java/javase/14/jpackage/packaging-tool-user-guide.pdf`

- *Docker docs*: `https://docs.docker.com/`

Index

A

abstract superclass
  versus interface 124, 125
Abstract Window Toolkit (AWT) 252
access control designation 108
adapter pattern 213-215
aggregation 132
annotation 112
Apache Foundation 36
Apache NetBeans 36
Applets 4
array data structure 156-158
ArrayDeque class 159, 160
ArrayList class 158, 159
association 130-132
autoboxing 84

B

Bean Validation
  used, for validating object managed by
    Context Dependency Injection 293
Bean Validation (BV) 290
behavioral pattern 215
BigDecimal class
  using 236-241

bill of materials (BOM) 241
binding 266
block 136
Boolean data type 73, 74
build section 56
  goals 57
  phases 57
  plugins 57-59
bytecode 5, 27

C

C 5
C++ 5
Cascading Style Sheet (CSS) 262-264
casting 80, 81
char data type 74, 75
checked exception 146
class 21
class composition 130
  aggregation 132
  association 130-132
classes 91, 92
  compound interest program, revising 94-97
  constructor method 93
  finalize method 92
  organization, based on functionality 98

class fields 88
    categories 88
class fields, categories
    class variables 88
    instance variables 88
class interface 121-124
    sealed class 125
    sealed interface 125
class-method region 91
class organization, based on functionality
    business class 100
    data class 98-100
    user interface class 101-106
classpath 7, 45
CMS Garbage Collector 5
code block 136, 137
code completion 33
coding structures 136
    code blocks 136, 137
    decision-making syntax 142-146
    expressions 138
    iteration 140
    operators 138-140
    statements 138
Collection interface 160
    streams, using 172, 173
Collections Framework 158
    Generics 161, 162
Collections Framework map structures 166
    HashMap 166-168
    LinkedHashMap 168, 169
    TreeMap 169, 170
command line
    Maven, running on 60-63
compile scope 56
constant 78
constructor method 93
Context Dependency Injection (CDI) 289

creational pattern 208, 211
custom JRE
    creating, with jlink 310

D

daemon thread 187
deadlock condition
    in threads 188, 192-196
decision-making syntax 142-146
default constructor 93
dependency 55
dependency inversion principle 205-208
dependency management, Maven 44, 45
Docker container system
    using 312
Docker Desktop system
    reference link 312
Docker image
    creating 314, 315
    working with 312, 313
documentation 221
    comments, adding 222, 223
    creating 222
    Javadocs 223-228
Domain-Specific Language (DSL) 41
do/while loop 141

E

Eclipse Foundation 35, 36
enterprise archive 53
exception classes
    creating 151, 152
exceptions handling 113, 146-149
    exception classes, creating 151, 152
    finally block 150
    program, ending 149

throws statement 150
throw statement 150
ExecutorService
thread pool, creating with 180-182
expression 138
Expression Language, for rendering pages
using 297-300
extension 118
extension inheritance 119

F

Facelets 297
Facelets, for rendering pages
using 297-300
Faces application
configuring 290-293
Faces life cycle 301
apply request values 302
invoke application 302
process validation 302
render response 302
restore view 302
update model values 302
Faces web application
deploying 301
factory pattern 211, 212
feature releases 8
finalize method 92
finally block 150
financial calculator program design 252-254
internationalization (i18n) 254-256
floating-point overflow 82
floating-point underflow 83
floating point unit (FPU) 236
foreach loop 141
for loop 140

G

G1 Garbage Collector 5
garbage collection 92
garbage collector 5
generic parameter 114, 115
Generics
in Collections Framework 161, 162
sequential implementations and
interfaces, using 162-166
Git 34
GlassFish 7.0 272
downloading 272
download link 272
installing 272
running 274
GPLv2 7
Gradle 41
Green Project 4
Groovy 41

H

hashCode 120
HashMap 166-168
heap 91
HomeBrew 9
HotJava 4

I

IDE
Maven, running in 63
identifier
casting 80, 81
constant 78
floating-point overflow 82
floating-point underflow 83

integer overflow 82
math library 85
naming 76-78
operators 78
wrapper class 83, 84
IEEE standard binary floating-
point numbers 72
image publishing 315
immutability 126
inheritance 90, 115-118
Object class, defining 119-121
installer packaging
with jpackage 310, 311
integer overflow 82
integers 71
integrated development
environment (IDE) 19
integrated development environments 32
Apache NetBeans 36
build systems 34
code editor 33
debugging 34
Eclipse Foundation 35, 36
JetBrains IntelliJ IDEA 38, 39
Microsoft Visual Studio Code 37
profiling 34
selecting, considerations 39
server management 34
source control management 34
integration testing 235
IntelliJ IDEA 38
IntelliSense 33
interface
versus abstract superclass 124, 125
interface segregation principle 204, 205
internationalization (i18n) 254-256
iteration 140

iteration approaches
foreach loop 141
for loop 140
while and do/while loops 141

J

J2EE 5
Jakarta 112
Jakarta EE 269
Jakarta EE 10 libraries/services 271
Jakarta Faces 270
Jakarta Server Pages (JSP) 270, 289
jar tool 15
Java
distributions 6
features 5, 6
functions 170, 171
history 4
installing 9
licensing 7
REPL 22-26
version, selecting 7
versions, timeline 8
Java 2 5
Java 2 Enterprise Edition (J2EE) 269
Java application
assembling 15
packaging 15
Java application, server
designing, ways 270
role 270, 271
Java archive 53
Java, as admin
installation, verifying 10
installing, for Linux 9
installing, for macOS 9
installing, for Windows 9

Java, as non-admin

environment variables, configuring 11

environment variables,
    configuring for Linux 11

environment variables, configuring
    for macOS 12

environment variables, configuring
    for Windows 11

installation, verifying 12

installing, for Linux 11

installing, for macOS 11

installing, for Windows 10

JavaBean 98

javac 14

execute process 27-29

two-step compile 27-29

Java classes

documenting 15

Java classpath 44

Java Community Process (JCP) 6

Java Database Connectivity 56

javadoc page, ArrayList class

reference link 16

Javadocs 223, 226-228

javadoc tool 16

Java EE platform 5

Java Enterprise Edition (JEE) 269

JavaFX GUI framework

Application 262

BigDecimalTextField 265

binding 266, 267

controls 266

CSS style sheets 263, 264

JavaFX bean 264, 265

pane 262

PrimaryStage 262

Scene 262

using 261

Java GUIs

history 252

java/javaw

execute process 27-29

two-step compile 27-29

Java Microbenchmark Harness library 260

Java native threads 176

Callable interface, implementing 182-185

Runnable interface, implementing 179, 180

Thread class, extending 176-178

thread pool, creating with
    ExecutorService 180-182

Java package 89

Java Platform Module System (JPMS) 308

Java program

compiling 14, 28

executing 14

running 28

working with 20, 21

Java Runtime Edition (JRE) 12, 309

JavaServer Faces (JSF) 270

JavaServer Pages (JSP) 270

Java source code packages 46-50

Java Virtual Machine (JVM) 5, 27, 91, 157

javaw tool 14, 15

JDK 12

JetBrains IntelliJ IDEA 38, 39

Jframe class 257

jlink 15, 27

used, for creating custom JRE 310

jmod tool 15

jpackage

used, for packaging installer 310, 311

jpackage tool 15

Jpanel class 257, 258

JShell 22-26

jshell tool 16

JUnit 5 241

testing with 241-245

L

Launch Single-File Source-Code Programs 29
  for Linux 29-31
  for macOS 29-31
  for Shebang files 31, 32
  for Windows 29-31
length 76
LinkedHashMap 168, 169
LinkedList class 159
linking exception 7
Linux
  Launch Single-File Source-Code Programs 29-31
  Maven, installing for 43
Liskov substitution principle 203, 204
literal value 71
Log4j2 229, 233, 234
logging 221
  java.util.logging 229-233
  Log4j2 233, 234
  using 228
Long Term Support (LTS) 8
looping 140
lossless conversion 72
lossy conversion 72

M

macOS
  Launch Single-File Source-Code Programs 29-31
  Maven, installing for 44
math library 85
Maven 41
  compressed files 42
  download link 42
  installing, for Linux 43
  installing, for macOS 44
  installing, for Windows 42
  running 60
  running, in IDE 63
  running, on command line 60-63
  used, for configuring web project 274, 275
Maven Central
  URL 45
Maven functionality
  dependency management 44, 45
  Java source code packages 46-50
  overview 44
  plugins 45
  project layout 45, 46
memory leaks 5
Mercurial 34
message bundles 256
methods 108
  access control designation 108
  annotation 112
  exception handling 113
  generic parameter 114, 115
  naming 112
  non-static designation 109, 110
  override permission 110
  override required 111
  parameter variables 112
  return type 111, 112
  static designation 109, 110
  thread setting 114
Microsoft Visual Studio Code 37, 38
modular Java
  exploring 308, 309
module 308
multithreading 196
MySQL 34

N

Napkin 252
native thread 176
NetBeans 36
No-Fee Terms and Conditions (NFTC) 7
non-daemon thread 187
non-default constructor 93
non-static designation 109, 110
Notepad 32

O

Oak 4
Object 119
Object class
  defining 119-121
object managed by Context
        Dependency Injection
  calculations 296, 297
  creating 293
  FinanceBean 293
  scope 294-296
  validating, with Bean Validation 293
object-oriented (OO) languages 170
object-oriented programming (OOP) 21
observer pattern 215-218
offending method 150
OOP, access control 88
  package 89
  package specifier 90, 91
  private specifier 89, 90
  public specifier 89
open/closed principle 201-203
OpenJDK 6
OpenJDK Community TCK License
        Agreement (OCTLA) 6

operating systems (OSes) 307
operators 78, 138-140
override permission 110
override required 111

P

package 41
package specifier 90, 91
Parallel Garbage Collector 5
parameterized testing
  performing 246, 247
parameter variables 112
Payara 34
plain old Java object (POJO) 289
plugins, Maven 45
polymorphism 127-130
POM files 53, 54
pom.xml configuration file 50-52
pom.xml file
  modifications to 275, 276
PostgreSQL 34
primitive data types 68
  Boolean 73, 74
  char data type 74, 75
  integers 71, 72
  literal value 71
  point, floating 72, 73
  type safety 68-70
private specifier 89
project layout, Maven 45, 46
Project Object Model (POM) 45
protected specifier 90
provided scope 56
public specifier 89
PyCharm 70

Q

quality assurance (QA) 235
query string
    accessing, in request by servlet 283, 284

R

race condition
    in threads 188-192
Read-Eval-Print Loop (REPL)
    in Java 22-26
record class 126
reference 76
regular expressions (regex)
    pattern, matching with 259, 260
Render Response phase 302
replace method 259
REPL tool 16
return type 111, 112
runtime scope 56

S

scopes
    compiled scope 56
    provided scope 56
    runtime scope 56
    test scope 56
scope, types
    @ApplicationScoped 294
    @RequestScoped 294
    @SessionScoped 294
    @ViewScoped 294
sealed class 125
sealed interface 125
Separation of Concerns 114, 200

sequential implementations and interfaces
    using, with Generics 162-166
sequential structures
    ArrayDeque 159, 160
    ArrayList 158, 159
    collection, declaring 160
    Collection interface 160
    implementations and interfaces, using 158
    LinkedList 159
Serial Garbage Collector 5
servers 34
servlet 269, 277-280
    accessing, query string in request 283, 284
    coding 277-280
    data, storing 284, 285
    requesting 281
Shebang files 31
    Launch Single-File Source-
        Code Programs 31, 32
Single-File-Source-Code feature 47
single responsibility principle 199, 200
singleton pattern 199, 208-211
Size in bytes 76
software design patterns 208
    adapter 213-215
    factory 211, 212
    observer 215-218
    singleton 208-211
software design principles 199
SOLID software design principles 200
    dependency inversion 205-208
    interface segregation 204, 205
    Liskov substitution 203, 204
    open/closed 201-203
    separation of concerns/single
        responsibility 200
Spring 112
SQL queries 34

stack trace 149
Star7 4
statements 138
static designation 109, 110
streams
    using, in collections 172, 173
String 75, 76
string operator 79, 80
structural pattern 213
subclass 90, 115
Subversion 34
Sun Microsystems 4
superclass 90, 115
Swing 252
Swing GUI framework
    controls and panels 260
    document filter 259
    event handlers 258
    Jframe class 257
    Jpanel class 257, 258
    pattern, matching with regular
        expressions 259, 260
    using 256
Swing GUI library 5
switch statement 33

T

Technology Compatibility Kit (TCK) 6, 272
Temurin 7
testing
    with JUnit 5 241-245
test scope 56
Thread, methods
    join() 185, 186
    sleep() 185, 186
    yield() 185

thread pool
    creating, with ExecutorService 180-182
thread priority 187
threads
    daemon thread 187
    deadlock condition 188, 192-196
    managing 185, 186
    non-daemon thread 187
    race condition 188-192
throws statement 150
Tomcat 53
TreeMap 169, 170
type safety 68-70
TypeScript 37

U

Ubuntu 20.04.4 LTS
    directory structure 13
unboxing 84
unchecked exception 147
Unified Modeling Language (UML) 115
unit testing 235

V

vi 32
view declaration language (VDL) 289
virtual threads 196
    creating 197
Visual Studio Code (VS Code) 37

W

web archive 53
web project
    configuring, with Maven 274, 275
WebRunner 4

web.xml file
   used, for configuring deployment 285, 286
while loop 141
WildFly 34
Windows
   Launch Single-File Source-
      Code Programs 29-31
   Maven, installing for 42
Windows Enterprise 11 Ver 21H2
   directory structure 13
World Wide Web (WWW) 4
wrapper class 83, 84

X

XHTML for rendering pages
   using 297-300

Z

Z Garbage Collector 5

‹packt›

www.packtpub.com

Subscribe to our online digital library for full access to over 7,000 books and videos, as well as industry leading tools to help you plan your personal development and advance your career. For more information, please visit our website.

Why subscribe?

- Spend less time learning and more time coding with practical eBooks and Videos from over 4,000 industry professionals

- Improve your learning with Skill Plans built especially for you

- Get a free eBook or video every month

- Fully searchable for easy access to vital information

- Copy and paste, print, and bookmark content

Did you know that Packt offers eBook versions of every book published, with PDF and ePub files available? You can upgrade to the eBook version at packtpub.com and as a print book customer, you are entitled to a discount on the eBook copy. Get in touch with us at customercare@packtpub.com for more details.

At www.packtpub.com, you can also read a collection of free technical articles, sign up for a range of free newsletters, and receive exclusive discounts and offers on Packt books and eBooks.

Other Books You May Enjoy

If you enjoyed this book, you may be interested in these other books by Packt:

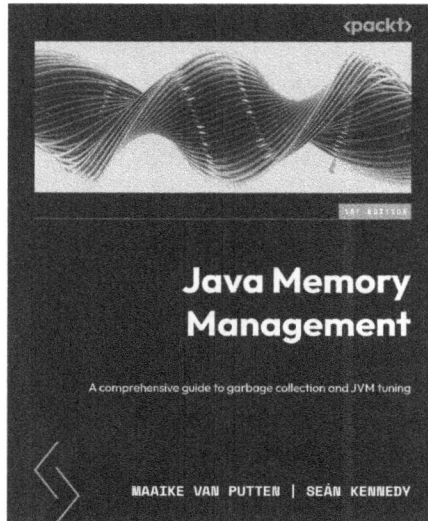

Java Memory Management

Maaike van Putten, Seán Kennedy

ISBN: 9781801812856

- Understand the schematics of debugging and how to design the application to perform well
- Discover how garbage collectors work
- Distinguish between various garbage collector implementations
- Identify the metrics required for analyzing application performance
- Configure and monitor JVM memory management
- Identify and solve memory leaks

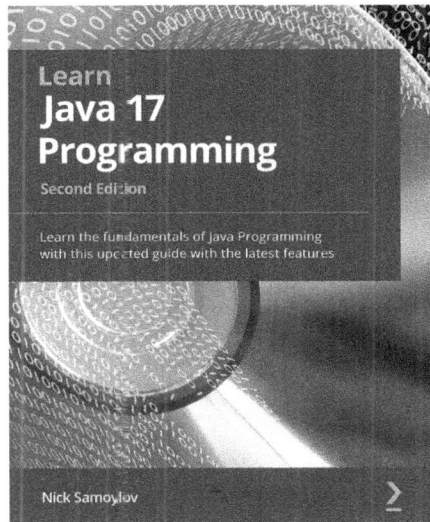

Learn Java 17 Programming - Second Edition

Nick Samoylov

ISBN: 9781803241432

- Understand and apply object-oriented principles in Java
- Explore Java design patterns and best practices to solve everyday problems
- Build user-friendly and attractive GUIs with ease
- Understand the usage of microservices with the help of practical examples
- Discover techniques and idioms for writing high-quality Java code
- Get to grips with the usage of data structures in Java

Packt is searching for authors like you

If you're interested in becoming an author for Packt, please visit authors.packtpub.com and apply today. We have worked with thousands of developers and tech professionals, just like you, to help them share their insight with the global tech community. You can make a general application, apply for a specific hot topic that we are recruiting an author for, or submit your own idea.

Share Your Thoughts

Now you've finished *Transitioning to Java*, we'd love to hear your thoughts! Scan the QR code below to go straight to the Amazon review page for this book and share your feedback or leave a review on the site that you purchased it from.

https://packt.link/r/1-804-61401-7

Your review is important to us and the tech community and will help us make sure we're delivering excellent quality content.

Download a free PDF copy of this book

Thanks for purchasing this book!

Do you like to read on the go but are unable to carry your print books everywhere? Is your eBook purchase not compatible with the device of your choice?

Don't worry, now with every Packt book you get a DRM-free PDF version of that book at no cost.

Read anywhere, any place, on any device. Search, copy, and paste code from your favorite technical books directly into your application.

The perks don't stop there, you can get exclusive access to discounts, newsletters, and great free content in your inbox daily

Follow these simple steps to get the benefits:

1. Scan the QR code or visit the link below

https://packt.link/free-ebook/9781804614013

2. Submit your proof of purchase
3. That's it! We'll send your free PDF and other benefits to your email directly

* 9 7 8 1 8 0 4 6 1 4 0 1 3 *